MISGUIDED WEAPONS

Related Titles from Brassey's

A Chain of Events: The Government Cover-up of the Black Hawk Incident and the Friendly-Fire Death of Lt. Laura Piper, by Joan L. Piper

Cold War Submarines: U.S. and Soviet Design and Construction, by Norman Polmar and K. J. Moore

Fire at Sea: The Tragedy of the Soviet Submarine Komsomolets, by D. A. Romanov

Silent Knights: Blowing the Whistle on Military Accidents and Their Cover-Ups, by Alan E. Diehl

USS Los Angeles*: The Navy's Venerable Airship and Aviation Technology,* by William F. Althoff

USS Ranger: *The Navy's First Flattop from Keel to Mast, 1934–1946,* by Robert J. Cressman

MISGUIDED WEAPONS

Technological Failure and Surprise on the Battlefield

Azriel Lorber

Brassey's, Inc.
Washington, D.C.

Library of Congress Cataloging-in-Publication Data

Lorber, Azriel, 1935–
Misguided weapons : technological failure and surprise on the battlefield / Azriel Lorber.—1st ed.
p. cm.
Includes bibliographical references and index.
ISBN 1-57488-395-X (alk. paper)
1. War. 2. Technology. 3. Military weapons. 4. Surprise (Military science) 5. Military art and science—History. 6. Military history, Modern. I. Title.
U21.2 .L674 2002
355.8—dc21 2002007381

Printed in the United States of America on acid-free paper that meets the American National Standards Institute Z39-48 Standard.

Brassey's, Inc.
22841 Quicksilver Drive
Dulles, Virginia 20166

First Edition

10 9 8 7 6 5 4 3 2 1

Qui desiderat pacem, praeparet bellum
(Let him who desires peace, prepare for war)

Vegetius

No one can guarantee success in war,
But only deserve it.

Winston Churchill

Dedicated to my family,
and particularly my wife, Nurit,
in appreciation of their patience
and understanding

CONTENTS

PREFACE

Consider the Battle of Crecy (1346), the Battle of Britain (1940), the first days of the Yom Kippur War (1973), and Operation Desert Storm (1991). Although separated by centuries and continents, is there a common denominator to all these battles? The answer is not readily obvious, but in all of these conflicts, the losing side—or at least the one in the most difficulty—did not fully appreciate the potential impact of perfectly known, even familiar, weapons in the hands of the adversary. Furthermore, the victims of this lack of foresight were not simply amateurs or citizen militias, but rather they were long-standing professionals who should have known better.

In describing this or that campaign or war, history books tend to describe the politics of the situation along with the thoughts and actions of the commanders and their troops. With few exceptions, personal attributes (such as wisdom, bravery, loyalty, and even piety) have commonly been considered the important and decisive factors, with discussions often neglecting to include the nature and quality of the weapons with which the combatants were equipped. This is rather strange because in the twentieth century at least, military technology is recognized as manifestly important and thus is the subject of countless books (including Bernard and Fawn Brodie's *From Crossbow to H-Bomb* and Martin van Creveld's *Technology and War*), articles, and scientific meetings held by such organizations as the American Defense Preparedness Association and the American Institute of Aeronautics and Astronautics. All of these publications discuss technological development, its relevance to tactics and strategy, political thinking and international relations, and even its social impact. Another important subject is its economic impact, both as a drain on society's resources and as a boon to the high-tech industry.

An important branch of this genre concerns books about military innovation, especially when that innovation happens to be based on some novel technology. Peter Rosen's *Winning the Next War* and Harvey Sapolsky's *The Polaris System Development* deal with organizational aspects, particularly the question of how a certain innovation manages to get into the mainstream of military thinking. But there are two problems with these discussions of military technology. First, most of this writing, whether concerned with concepts such as amphibious or armored warfare, or when dealing with hardware, such as aircraft carriers, submarines, radar, various aircraft, and the like, deals with big undertakings. This is understandable; an aircraft carrier or a supersonic airplane is much sexier than a snorkel or a few boatloads of chemicals, but wars, or at least campaigns, have been affected by both. Second, what happened on those occasions when the novel technology (or even a weapon system based on it) was already in existence, at least in the hands of one of the enemies? Did both sides treat it "properly," so to speak? Finally, do all military organizations identify major and critical innovations in military technology; and if so, do they strive to introduce them into the force structure, or develop countermeasures if they are in the hands of the adversary?

Some recently published books that describe and analyze military failures and bungled campaigns are also concerned almost exclusively with the actions or, to be blunt, the various forms of gross incompetence of commanders and their subordinates. With few exceptions, the role that particular weapons played in the success or failure of certain battles or campaigns has usually been neglected. Even when the role of the weapons in deciding the outcome has been acknowledged, such as radar during the Battle of Britain, rarely has an effort been made to examine the roots of the asymmetrical situation more deeply, to analyze how it came about, or to look at it from the other side of the barricade. In the above cases, why and how did the French, the Germans, the Israelis, and the Iraqis get into such predicaments? And this is of course true of many other wars. We know a lot, for example, about the exploits of the various commanders during the fighting in the Western Desert or the breakout from the Normandy beachhead. Little, however, is ever said about the relative merits of the tank guns on both sides. In both of these campaigns, the potentially disastrous outcome for the Allies was prevented by several other factors, mostly a preponderance of materiel and uncontested air superiority. Afterward such mistakes were swept under the rug and eventually faded, even from the history books. This oversight is often compounded by neglecting discussion of esoteric, but still important, technologies that somehow never made it into the limelight. One such example is the very slow introduction, by both sides, of auxiliary (drop) fuel tanks for fighter aircraft during the various stages of World War II.

While many, if not most, past battles and wars were indeed decided by the traditional "qualities" of the troops and their leaders, this book will deal with only one aspect of the problem: "technological failure." Technological failure, by my definition, involves the lack of comprehension of the effect that certain weapons, or the lack thereof, may have on the conduct of warfare. On a somewhat different level, a technological failure may also involve the lack of awareness of the science and technology involved in a particular weapon. Finally, technological failure is concerned with people and their attitudes toward the ever-changing world of technology. I wish to emphasize that in the present context, the term "technological failure" does not refer to engineering failure, poor design or workmanship, mechanical breakdowns, shoddy maintenance, and similar occurrences that sometimes apply to whole classes of weapons.

Being an engineer, it seemed to me that at least some of the "technological failures" on the battlefield had to result in technological surprises, leading to disasters for the surprised. However, I had never encountered the term "technological failure" in the literature, and only once the term "technological surprise" in the meaning I assign it. Upon further reflection, it became clear that there were many military defeats that in one way or another were caused by inferior weapons (or their total absence), but only rarely was this lack or inferiority the fault of the weapon designers. Usually it was a lack of knowledge, or the total misunderstanding, by all sorts of "higher-ups," of technological and scientific facts and their application to warfare. It was also clear to me that at the time of the said defeats, and in many cases even much later, this fact was not truly appreciated.

This book is not a text on military history as such, nor is it a technical text. It is an attempt to present some of the more important, sometimes decisive, technological failures of the past. In order to offer better understanding of these failures, it was necessary on occasion to provide a somewhat wider background and its relevance to the problem. The historical and anecdotal aspects of past technological failures and surprises are fascinating to read about. With twenty-twenty hindsight, we delight in reading and marveling at the shortsightedness, or downright stupidity, of the military commanders and their political leaders who let themselves be drawn into these difficulties. But as with most misfortunes, there are lessons to be learned and applied, even in our "enlightened" times, and the two most pertinent to our discussion are the following questions: Can a technological failure—or for that matter, a technological surprise—still happen today when everybody, or at least those in responsible positions, is aware of the destructive potential of such an event? And, is there a way to guard against such occurrences?

The answer to the first question is obviously "yes." One case in point, which will be elaborated upon later, was the inability of the Patriot batteries to intercept the Iraqi Scuds in the Persian Gulf War of 1991. This failure could have been predicted years earlier. The necessary information was there for anybody to use, except that somehow it did not register. As to the second question, I do not believe that there is a surefire solution to the problem of how to guard against technological failure. But being aware of the problem is a large step in the right direction. Considering the human race's unbounded capability to do silly things, military commanders and their staffs, if they bothered to learn anything from the mistakes of their predecessors, should be aware that technological failures and technological surprises could happen at any time and age, including our own, or in the future. In too many cases technological failures and surprises stem from too human characteristics such as self-satisfaction, disdain for the adversary, obtuseness, and conservatism, or in other words, stupidity and lack of professionalism. Unfortunately, these traits are not going to go away by edict.

One commonly encountered question is, "Why discuss failures?" It is true that describing successes may be morally more uplifting. Also, in most cases, one's (or one's army's) failure is the other side's success. But the main reason for discussing failures is simple and has been known for a long time. You learn much more from failures—mostly yours—but if properly studied, also from the failures of others. To paraphrase Samuel Johnson, failure concentrates the mind wonderfully.

Most of the cases presented in this text deal with World Wars I and II. These were the first truly technological and industrial wars and as such are rich in illustrative examples. Not all the failures were earthshaking, monumental ones. Many started in a minor fashion, but as we all are aware, "for the lack of a nail, a kingdom was lost." There will be some brief mention of technological intelligence failures, in which a device was concocted by the "back room boys" and then unleashed on an unsuspecting adversary. These, however, are not true technological failures in the present meaning, unless knowingly ignored. After all is said and done, such failures are not as unpardonable as those failures where the information was available, or discovered the hard way, but nothing was done about it.

A few words about references. Since this is intended to be a popular book, I did not want to burden the reader with "proofs" to the most basic and generally accepted facts. Consequently, I refrained from quoting endless references, and furthermore, I felt free to use secondary sources. I did provide references when certain authorities are the only source for information, quoting previous authors, or in cases that are very important or bizarre. I also tried to specify the sources whenever I say less than gracious things about individuals or organizations. However, I used general information from all the sources in the bibliography, even when they are not specifically cited, and I acknowledge my debt to all these authors for endless hours of enjoyable reading and for their (occasionally subtle) contribution to the present volume.

I wish to thank the following people who in various ways

helped bring this work to conclusion. First and foremost, Professor Harvey Sapolsky, Director of the Security Studies Program at MIT, and Professor Ted Postol at the Security Studies Program, who enabled and helped me to spend my sabbatical (during which time this book was written) with this program. I wish to thank Dr. Dennis M. Bushnell, Chief Scientist of NASA's Langley Research Center; and Dr. Owen R. Cote Jr., Associate Director of the Security Studies Program at MIT, who read the manuscript and made many useful suggestions. The same goes for Michael Eisenstadt of the Washington Institute for Near East Policy, who also provided a critical reference to one of the cases described. All mistakes and inaccuracies that remained after these gentlemen's comments are mine alone. I would like to thank the anonymous reviewer of this book, who pointed out to me a very interesting reference that made one of the cases described much stronger. I also wish to thank Dr. Benjamin Frankel, editor of the journal *Security Studies,* for his help and advice and Dan Kronenberg, publisher of Kronenberg Professional Books for his permission to use several excerpts from my last book, published under that imprint. I would like to thank George Maerz, from the National Defense University (NDU) Press, and other NDU library personnel for their help in obtaining hard-to-get material. I would also like to thank the MIT libraries' staff for their help in making my research possible, easy, and fruitful. Finally, I would like to thank Rick Russell and Paul Merzlak at Brassey's, Inc. for all of their exceptionally well-grounded comments, suggestions, patience, and help with this project.

Prologue

Early Technological Failure: The Battle of Crecy

Adherence to dogmas has destroyed more armies and cost more battles than anything in war.

J. F. C. Fuller

The Battle of Crecy was part of the Hundred Years' War between the English and French kings over the title to some estates in France. In one of the early engagements of this war, in the spring of 1346, Edward III, the English king, landed on the French coast and went on a lengthy raid through the French king's land. King Philip VI went out to confront him, and the British withdrew until, exhausted, they stopped near the village of Crecy on the 26th of August. The English forces comprised some four thousand knights, six thousand Welsh archers equipped with longbows, and about five thousand Welsh infantry. The French mustered twelve thousand knights, six thousand Genoese crossbowmen, and some twenty thousand militias. The exact numbers differ from source to source, but it is generally accepted that the force ratio was better than two to one in favor of the French.

King Philip VI, who was in pursuit of King Edward in the early afternoon of August 26, actually stumbled across the English, who had positioned themselves on a slight hill with a wood on one flank and a river on the other. Most of the English knights had also dismounted. Apart from their role as heavy infantry, dismounting was a clear signal to the infantry that nobody was to run away if the situation got sticky.

The French king wanted to postpone the attack until his entire force, stretched along several miles, came together. Philip's knights, however, urged him to act immediately, and he sent the Genoese crossbowmen forward. Their bolts failed to reach the English line, and under heavy fire from the Welsh bowmen, who were firing downhill, the Genoese forces retreated. Seeing this, the French knights charged up the hill, trampling the crossbowmen under the hooves of their horses, but were stopped short of the English positions by a hail of arrows that caused severe losses. Through that long summer day and evening, the French charged up that hill sixteen times, and although on some occasions they came in contact with the English, they did not penetrate the line. The French finally gave up near midnight, leaving more than fifteen hundred knights and some ten thousand others dead. The English lost three knights and about a hundred infantrymen. It is important to point out that this discrepancy in casualties should be considered in view of the original force ratio.

The English victory in the Battle of Crecy had some long-standing consequences. It enabled King Edward to lay siege to Calais, which eventually fell and remained in English hands until 1558, long after the end of the Hundred Years' War. It advanced England to the status of an international power and further enhanced the reputation of the longbow. Finally, the Battle of Crecy was probably the first time since the battle of Adrianople (A.D. 378, in which a Gothic cavalry force annihilated Eastern Roman infantry legions) that a force predominantly comprising infantry stopped a numerically superior force of cavalry in its tracks and won the day.

During the Hundred Years' War there were many more campaigns, and in two of them the French again tangled with the longbowmen, with similar results. These were the Battle of Poitiers, in 1356, and the Battle of Agincourt, in 1415. But the seeds of these French disasters were sown much earlier than the Battle of Crecy, which in itself became a symbol, and the lessons of those earlier encounters are really the moral of this story.

On November 11 (St. Martin's Day), 1337, the Earl of Darby, in the service of Edward III, landed on the coast of Flanders in what came to be regarded as the opening battle of the Hundred Years' War.[1] Darby was opposed by Flemish crossbowmen who lined the port's quays. The crossbowmen were driven away under the heavier fire of the Welsh longbowmen, and the landing was successful. However, the French apparently did not assimilate the results of this battle. Pending further operations against France, Edward III realized that he would have to possess naval control of the channel so that he would be able to travel unhindered back and forth. On June 24, 1340, an English fleet of some 250 ships attacked a combined French-Spanish fleet (under French command) of about 200 ships at anchor in Sluis (or Sluys, on today's Belgian-Dutch border). Volleys of arrows fired from the English ships opened the engagement and decimated the French and Spanish crews. The battle ended with an almost total destruction of the combined continental fleet and the loss of some twenty-five thousand men, compared with English losses of some four thousand.

No matter how militarily inefficient the feudal system of the fourteenth century was, it is certain that accounts of battles did survive for six years, and somebody should have paid attention to them. It is known that the disaster at Sluis was reported to the French king, but it appears that the French leadership took no heed of the details.[2] The only explanation for such recurrent debacles is that the knights of that period had some kind of taboo about discussing lost battles and reasons for the defeat. Historically, of course, this assumption is totally unsupported, but on the other hand, such monumental stupidity does need some explanation. Is there a better one?

Surprisingly, the longbow turned out to be the weapon of decision. Almost two meters long and firing a "yard-long" arrow, the longbow was developed in Wales as a hunting weapon. Many successful weapons were used in a "dual use" role, both for warfare and for hunting, and the longbow was no exception. It was made of elm, hazel, yew, or ash, and if made properly the natural wood branch from which it was made contained elements that worked in both tension and compression. A similar principle was employed in the Asiatic bows that were made of wood, horn, and animal skins, and that today we will define as made of composite materials, except that the Welsh bow was much simpler. This bow was similar in effectiveness to the crossbow but had a rate of fire four to five times higher. The catch was that it required lengthy and constant practice to achieve and maintain proficiency. Thus, the longbow pitted the "price" of a highly trained professional against the price of a mechanical contrivance, the crossbow, which could be used by a soldier of considerably less training. But Wales was poor, and so the equation leaned heavily in favor of the longbow that superseded the crossbow in England.[3] Welsh longbowmen were used extensively in the English wars against the Scots, and the bow acquired a fearsome reputation, albeit a local one. The French, however, even after Sluis, apparently were not impressed.

They maintained this attitude. After losing the Battle of Crecy, ten years later they lost the Battle of Poitiers and then the Battle of Agincourt. In the latter two engagements, considerably larger French forces (at Agincourt it was fifty thousand French against six thousand English) attacked well-prepared English positions and suffered terrible losses, mostly to the archers. Admittedly, in both these engagements the French dismounted some of their knights to attack on foot (thus avoiding the confusion and damage caused by wounded horses), but to no avail.

Realizing that the enemy was doing something different, the French attempted a half-hearted doctrinal change (dismounting the knights) to overcome a consistent tactical inferiority. After Crecy they even considered adopting the longbow, but gave it up when they discovered that they did not have the required pool of trained people. But still they persisted in *l'offensive à outrance* ("charge at all costs," although this term was not coined until centuries later) regardless of the consequences.

These battles took place before the word "technology" was invented, but they illustrate very well the meaning of technological failure, which in this book includes the complete disregard, for whatever reason, of the effect that a different, often novel technology exerts on a given battlefield.

1 Predicting the Outcome of Conflicts

You know well how you are entering a war. You never know how you are going to emerge from it.

Unknown

The race is not always to the swift, nor the battle to the strong.

Ecclesiastes 9:11

INTRODUCTION

Throughout history disagreements on political, religious, economic, and territorial matters or ones concerning honor, face, or plain megalomania have erupted into armed conflict. Often leaders entered such conflicts without much previous thought or planning, although the outcome of such wars was of major interest and importance to all involved. The uncertainties of such situations and their importance to the participants were already described in the Bible (I Kings 20:11). There, King Ahab mocks the bragging Ben-Hadad, king of Syria, by saying, "Tell him, let not him that girdeth on his harness boast himself as he that putteth it off." Ben-Hadad's bragging brings to mind another more recent boast, about "the mother of all battles."

For those directly involved it was literally a question of life or death, the curtailment of freedom (or even outright slavery), financial losses, humiliation, and many other kinds of hardship. For others it could mean changes in government, new taxes, or even oppression and looting. Naturally, everybody involved wanted to predict accurately the outcome of a war or a battle, preferably even before it took place, or before everybody else knew about it.

THE IMPORTANCE OF QUANTITATIVE FACTORS

What criteria did the professionals and the soothsayers use to predict the outcome of armed conflict? Foremost, these criteria involved such tangible factors as the number of enemy soldiers and information about their weapons. Intuitively, and to a large extent, historic experience bears this out. One might predict that bigger armies have a better chance of winning. Napoleon expressed this sentiment best by saying that "God sides with the bigger battalions." Even today a simple count (or comparison) of the numbers of divisions, tanks, and aircraft is expected to give an indication of an army's chances in a real conflict. This is why all sorts of lists of the "order of battle" of the various countries of the world are published annually and successfully marketed.

Herein lies a problem as old as warfare. There is magic in large numbers and a temptation to obtain quick results by simple arithmetical means. Occasionally, however, when such predictions are based on raw numbers, the actual results can be surprising, if not downright embarrassing, for the "experts." In May of 1948, the accepted wisdom of the world's leading military authorities gave the newly established state of Israel less than two weeks against the onslaught of the combined Arab powers. The Israeli military at the time was a ragtag collection of local militias, who were essentially without formal military training, modern combat planes, and artillery, and who had only some homemade armored cars and two stolen British Cromwell tanks, but without spares. Facing the Israelis were six organized armies (Egypt, Sudan, Jordan, Lebanon, Syria, and Iraq) with all the trappings of regular armies, including armor, artillery, and real air forces and navies, and supported by large local guerilla forces. The results are a matter of historic record, with scholarly papers still arguing what went wrong for the Arab forces.

In 1991, before the second Gulf War, many of a new generation of authorities predicted a drawn-out bloody campaign against the enormous "battle-hardened" Iraqi army, involving thousands of coalition (mostly United States) casualties. It was feared that the well-equipped Iraqi army would prove worthy of its equipment and its earlier "battle experience" against the Iranians in the first Gulf War. One researcher even made comparisons with the Arab-Israeli wars and predicted hundreds of Allied killed, even under the best of circumstances, including total Allied mastery in the air. How this prediction failed will be discussed later in this text. For now, suffice it to say that a full list of such miscalculations for the last few centuries would probably take several pages, but it should be apparent by now that numbers alone do not tell the full story.

THE QUALITY OF TROOPS AND COMMANDERS

The root cause of such mistakes lies in neglecting the well-known fact that the size of an army's forces is not the only deciding factor in whether a battle is won or lost. There are other factors influencing the fighting capability of the armed forces. These are the intangibles that create and sustain the overall quality of the fighting force. They range from the basic quality of the manpower (meaning education, intelligence, and resourcefulness) and their morale and belief in the cause, to the quality of their weapons.

The quality of the troops, as a military unit, also depends on their level of training and discipline, their previous battle experience, and their confidence in their commanders and in themselves. This last factor is particularly difficult to quantify except in the most general terms. Leo Tolstoy, the famous Russian novelist (1828–1910), saw action at the Siege of Sevastopol during the Crimean War (1853–1856). He later made a very profound observation about the will to fight: "[T]he best plans of the high command will come to naught if at the crucial moment Ivan will refuse to get out and charge the enemy."

An even more important consideration is the ability—and in too many cases, the lack of ability, or incompetence—of the higher command. A brief review of military history will show that more battles ended as they did, not because of the brilliance of the winning commander, but rather because of the inability of the losing one. The defeat of the Spanish Armada, Custer's infamous last stand, and the Gallipoli campaign of 1915 fit this category. Such incompetence is reflected not only in plain misunderstanding of the tactical situation or cowardice, but more often in such mundane problems as timely payments to the troops, proper supplies, and attention to sanitation. Until the twentieth century, more soldiers on campaigns perished because of poor health and sickness than due to enemy action. Furthermore, the quality of the commanders at all levels is very difficult to define and even more difficult to quantify. First, this quality consists of a large number of subconstituents, whose interactions are not always consistent, and on occasion may in fact be contradictory. The difference between daring and recklessness is measured only by the end result, and compassion can detract from efficiency. Second, how does one quantify all these subtle traits of resourcefulness, audacity, bravery, and even compassion? Finally, because of the burdens of command, occasionally combined with advancing age or failing health, the performance of leaders under the pressure of combat may not be consistent over the years. This results in another unpredictable factor, which is impossible to foresee, no matter the man's previous record. It is generally accepted that Napoleon's later performance suffered because of deteriorating health, and combined with his enemies' improved capabilities, culminated in his final defeat.

OTHER FACTORS, INCLUDING COMMON SENSE

There are two other factors that do not seem to be directly related to quality or competence but that are considered important. The first is luck. Bad luck too often has been blamed as the cause of misfortunes in warfare, but careful attention to details could, and still can, provide against most cases of "bad luck." After all, "chance favors the prepared ones." But traditions die hard. Even in modern war gaming, "luck" is often represented by a random throw of dice or a computerized version of it.

There has often been another important consideration—the

belief of many in the power of celestial intervention, or at least celestial guidance. The Old Testament relates how "they fought from heaven; the stars in their courses fought against Sisera" (Judges 5:20). Stars or no stars, it is obvious that that particular defeat occurred when Sisera, in response to Barak's presence on Mount Tabor, took his heavy chariots (a technological asset the Israelites did not have and could hardly fight against) into the Kishon valley in the middle of the winter. This is a marshy terrain in all seasons, and thus Sisera got bogged in and lost his advantage—and subsequently his head. We do not have the tools to determine whether this happened because of poor intelligence about the terrain (rather doubtful since Sisera had his headquarters a few miles from there) or because of sheer arrogance and disdain of the facts of geography and weather. Either way, this incident is worth noting because here we are back to another basic quality necessary for higher command—common sense.

The lack of common sense, leading to irresponsible behavior, was not limited to absolute or ignorant rulers of antiquity who used to execute anybody who differed with them in the councils of war. This kind of thoughtless approach can be abundantly found in modern times, even in the enlightened democracies, and achieved its nadir during the First World War.[1] The capabilities of the magazine-equipped rifle and the modern machine gun were known fairly well by the time of the Second Boer War (1899–1902), and the combination of these two weapons proved quite deadly during the Russo-Japanese War in 1904. Nevertheless, the French generals truly believed in the power of the attack and were convinced that élan would prove a match to the machine gun. The Germans, who found themselves behind static lines after their flanking moves were countered, may not have believed so much in élan. However, disregarding previous lessons and having no better solutions, both sides simply fell back on their staff training. To give the Germans credit, though, in 1915 they tried gas (first chlorine and then other compositions) but mismanaged that too. After almost three years, several million casualties, and facing numerous mutinies in the trenches, both sides started to look for better ways.

QUALITY VS. QUANTITY

It is simple to predict the outcome of an armed conflict when both sides are of more or less similar quality but with one side sufficiently larger to ensure victory. It is also easy to predict the results when similarly sized armies, but of markedly differing quality, face off.[2] The difficulties rise when a small but higher-"quality" army confronts a larger one of dubious or average quality. These predictions, of course, are possible primarily in terms of a single campaign or a very short war. If the two opposing countries differ much in size and overall resources, it can be assumed that unless the smaller army achieves a quick victory (or settlement of the conflict), if the national resources of the bigger side can be properly mobilized and organized, it will eventually prevail should the war continue. It should be noted that the mobilization of resources, the organizational aspects, and the will to fight on are of paramount importance. When lacking, or if the conflict is geographically too far afield, the "weaker" side may eventually win.

There have been several conflicts in which smaller military forces won against potentially overwhelming odds. Consider the American War of Independence, the Russo-Japanese, and the Vietnam Wars against the French and the Americans. Another example is the war between Prussia and Austria (1866). The Austrian Empire was what we would call today a "paper tiger"—huge, but poorly organized—with its military led by a sixty-two-year-old general with an excellent record (he saved the day in Solferino in 1859), but who by 1866 was really too ill for the position, and because he was not familiar with the terrain accepted the command very reluctantly. The decisive battle at Sadowa was fought between equal forces, some two hundred thousand on each side, and ended with a crushing Austrian defeat. It is generally accepted that this victory was largely due to a novel breech-loading rifle (the Dreyse needle gun invented by Nikolaus von Dreyse), with which the Prussian infantry was equipped. Admittedly, the rifle was not yet perfect. Its breech leaked hot gases (because it had a paper cartridge), and its range was about half that of the standard muzzle-loaders of the time, but it could be fired about six times a minute, compared with two times per minute for muzzle-loaders. Additionally, this rifle enabled the Prussians to fire from the prone position, while a soldier using a muzzle-loader had to fight, or at least load, while standing up, presenting a much bigger target. Interestingly, after being developed around 1838, the rifle was immediately ordered by the Prussian army, and the Dreyse plant was busy for some ten years supplying the Prussians with the initial and follow-up orders. In 1851 the rifle was offered to the Austrians but turned down. The Austrians claimed that "although the needle gun permits rapid fire as long as there is no stoppage, this does not constitute any real advantage, because rapid fire will merely exhaust the ammunition supply." But there was another reason the Austrians decided to decline the needle gun. The Austrian manufacturing plant for the old rifles had just finished retooling for more efficient production.[3] Adopting a completely new rifle would have been a financial calamity. Only just before the war did the Austrian military realize the situation and make efforts to obtain better guns, but by then it was too late, and Prussia wrested from Austria the hegemony in central Europe.

Mountains of paper and rivers of ink were expended in arguing about quantity versus quality of the armed forces and on the relative merits of the various qualitative attributes and how to enhance them. Before the twentieth century, most of the writers in this field were military officers of varying ranks, with various degrees of experience. Some of the better-known ones are Maurice De Saxe (1696–1750); Frederick II (the Great, 1713–1786); Karl von Clausewitz (1780–1831); Antoine Jomini (1779–1861); Ardant Du Picq (1821–1870); Ferdinand Foch (1851–1929); and from far away, Sun Tzu (c. 500 B.C.).[4]

Clausewitz is the most famous of these and probably had the most influence on military thought, although as will be shown later, in some aspects this influence was detrimental. He was a Prussian staff officer during the Napoleonic Wars. Although he never actually commanded troops in combat, he wrote one of the most exhaustive books on the art and science of warfare. Clausewitz died of cholera in 1831, and his books were edited and published by his widow. Clausewitz discussed the whole gamut of subjects dealing with management of troops and campaigns, and among others he developed and codified the concept of the "principles of war." These include perseverance, unity of command, fighting spirit, initiative, concentration of effort, surprise, and security. These principles are taught even today in military academies all over the world. While one may argue with some of his dictums (concerning information [intelligence] in war: "[A] great part of the information obtained in war is contradictory, a still greater part is false and by far the greatest part is of a doubtful character"), they should be understood in the light of his era.[5] On the whole, there is no doubt that most of Clausewitz's teachings are still applicable today, having stood the test of time.[6]

Although these and many other prominent military leaders before and after dealt with the problems of quality, they all focused on the quality of the troops and commanders. Such quality was achieved by a most rigid discipline and endless, rigorous training. De Saxe, for example, first became famous for the musketry training of his regiment.

LANCHESTER'S EQUATIONS

Not until the First World War was a scientifically based attempt made to quantify the problem of quality versus quantity in military operations. This type of attempt was first made by William F. Lanchester, a British engineer who worked in many fields. Lanchester, however, is remembered for essentially inventing the discipline of operational research in a paper he wrote in 1916 titled "Aircraft in Warfare; the Dawn of the Fourth Arm." Lanchester developed the so-called Linear Law and the Square Law, both named after him. Basically these laws are differential equations that describe the attrition (killing) rates of two opposing forces of infantry of different quality and size.

The results of this novel way of treating an age-old problem were surprising to say the least. The original work was largely concerned with the question of concentration of effort. It came to the (numerically backed) conclusion that the maximum quantity (numbers) of available resources should be committed (space permitting) concurrently, instead of piecemeal. In this concept, Lanchester proved mathematically what is considered one of the important factors that contributed to Napoleon's victories, and that Clausewitz later formulated as one of his "principles of war"—the concentration of effort. Later, during the Battle of Britain, legless fighter pilot Douglas Bader empirically (presumably) came to the same conclusion when he advocated the use of the "Big Wings" of defending fighter aircraft. But the most important and most surprising aspect of Lanchester's work was that it has shown, in a mathematically rigorous way, that under certain circumstances quantity is more important than quality. In other words, if you can invest a certain amount of resources in increasing quality, by say 10 percent, or in increasing quantity by the same amount, you will do much better by investing in quantity.

It should be remembered, however, that Lanchester's work should not be applied beyond its limits. Obviously, almost any real military operation is considerably more complicated than the mathematical models could handle. In other words, the equations could be used only for very simple scenarios. And, Lanchester's work ran into the problem of accurately quantifying the various aspects of quality. Quality is not a single entity, but consists in itself of scores, if not hundreds of "subqualities" that widely range in character and value. These, then, must be summed up in some way, after being assigned a numerical value for themselves and after some weighting function as to their importance in the overall scheme of things is devised and applied.[7]

Furthermore, this kind of calculation assumes that the qualities of the two adversaries, while differing quantitatively, are still of a similar nature. Experience, however, has taught us time and again that this is not always the case. How do we compare the merits of the tank and the antitank missile, or how do we define the "quality" of a suicide warrior, such as a kamikaze pilot or a suicidal fanatic, with a bomb? Still, such an analysis can be useful by discerning trends, which can be affected by trade-offs, for example in the design of weapons.

It is surprising that in all the discussions of the quality of military forces, including by mathematical analyses, the effect of technological innovation is missing. By technological innovation we do not mean an incremental improvement in an existing weapon, but one military side's introduction of something totally new that causes, mathematically speaking, a discontinuity. What happens when an absolutely new, possibly revolutionary piece of equipment makes its debut in the beginning or the middle of a shooting war, and what might be its effect on the outcome? But there is also the obverse side. What might happen if such novel equipment is available but one side cannot see its potential? Some examples of such cases will be examined in the following chapters.

2 The Roots of Technological Failure

Nothing so comforts the military mind as the maxim of a great but dead general.

Barbara W. Tuchman

THE ROLE OF WEAPONS

The previous chapter mentioned some of the writers who dealt with the various aspects of the quality of the troops and their officers. Surprisingly, none of these writers wrote about the possibility of improving the quality of the total fighting force (as different from the troops) by materially improving the weapons with which these troops so laboriously trained. Nor for that matter do the works of the classical military writers contain any discussion of the importance of weapons, even contemporary and conventional, and their relative merits in attaining victory. The importance of the correct use of extant weapons for waging a successful war was well recognized, because discussions of the correct and efficient use of artillery and musketry and of training in these weapons are quite common. Here again the emphasis is on the quality of the soldiers in operating given weapons and not the quality of the weapons or the possibility of improving them. Admittedly, the period's style of warfare and pace of weapons development had a lot to do with this attitude. Thus it was accepted that weapons are at best of secondary importance. Clausewitz in fact said it very clearly: "The inventions have been from the first weapons and equipment for the individual combatants. These have to be provided and the use of them learnt before the war begins. They are made suitable to the nature of the fighting, consequently are ruled by it; but plainly the activity engaged in these appliances is a different thing from the fight itself; it is only the preparation for the combat, not the conduct of the same. *That arming and equipping are not essential to the conception of fighting is plain* [italics added], because mere wrestling is also fighting."[1]

In some respects this is bad enough, but trust Clausewitz to make his meaning clear, and so he goes on to emphasize:

Two Modes of Fighting—
Close Combat and Fire Combat

(46) Of all weapons which have yet been invented by human ingenuity, those which bring the combatants into closest contact, those which are nearest to the pugilistic encounter, are the most natural, and correspond with most instinct. The dagger and the battle-axe are more so than the lance, the javelin, or the sling.

(48) Although there are shades of difference, still all modern weapons may be placed under one or other of two great classes, that is, the cut-and-thrust weapons, and fire arms; the former for close combat, the latter for fighting at a distance.

(50) Both have for their object the destruction of the enemy.

(51) In close combat this effect is quite certain; in the combat with fire-arms it is only more or less probable. From this difference follows a very different significance in the two modes of fighting.

(52) As the destruction in hand-to-hand fighting is inevitable, the smallest superiority either through advantages or in courage is decisive, and the party at a disadvantage, or inferior in courage, tries to escape the danger by flight.[2]

THE ORIGINS OF VARIOUS MISGUIDED ATTITUDES

No doubt this counsel was good for Clausewitz's time, and probably he did not realize the future impact of his writing. However, it is easy to see where future generations of military leaders got their notions about the importance of the playing fields of Eton (for the victory at Waterloo) and the superiority of élan over the power of concentrated machine gun fire. This kind of thinking, coming straight from the guru of the art of war, acquired a life of its own and then led to further utterances, two of which are quoted below:

> It must be accepted as a principle that the rifle, effective as it is, cannot replace the effect produced by the speed of the horse, the magnetism of the charge and the terror of cold steel (British cavalry training manual, 1907).

This was written fifty-three years after the disastrous charge of the Light Brigade at Balaklava, in which from a force of about seven hundred, more than five hundred were casualties, and for no purpose at all.[3]

The second quote is even more illuminating:

> We would rather have a classically educated boy than one who has given his mind very much up to electricity and physics and those kind of subjects. Power of command and habits of leadership are not learned in the laboratory. Our great point is Character; we care more about that than subjects (Lieutenant Colonel Murray, Assistant Commandant at Woolwich, 1902).[4]

While these authors dealt with land warfare, this approach apparently had its adherents on the sea as well. In 1898 the British navy started thinking about the all-big-gun battleship. An Italian ship designer, Vittorio Cuniberti, advanced the original concept that if all the guns of a battleship were of one caliber (the biggest possible), fire control would be simpler and accuracy improved. It is almost tempting to say "naturally," but the design was rejected by the Italian Admiralty, picked up by the British, and eventually Dreadnought battleships grew out of it. Rear Admiral Alfred Thayer Mahan (1840–1914) of the U.S. Navy, "The Clausewitz of Naval Strategy," opposed this concept "because such vessels would fight only at great ranges. These ranges would create in the sailor the indisposition to close. They would thus undermine the physical and moral courage of a commander."[5] Mahan died in

1914. Ranges were opening up before 1905. In spite of his thinking, bigger and bigger battleships were built, with greater ranges, and the aircraft carriers finally put hundreds of miles between opposing forces.

The end result was that from the end of the eighteenth century and until the end of the First World War these attitudes lingered, in some quarters even until the Second World War, with complete disregard for the accelerating pace in weapons development and the changing character of war.[6] "In the teeth of technological revolution, as it were, Europe's soldiers sought to wage a future war by the methods of the past rather than of the age to come."[7] Not to be outdone by the Europeans, the situation in the U.S. was hardly better. "Internal barriers to change and the myopic vision of single-issue constituents contributed significantly to the Army's lack of preparedness for World War II."[8]

To be fair, there were other voices. Major General J. F. C. Fuller, an important writer on military matters, wrote that "tools, or weapons, if only the right ones can be discovered, form ninety-nine percent of victory. . . . Strategy, command, leadership, courage, discipline, supply, organization and all the moral and physical paraphernalia of war are nothing to a high superiority of weapons—at most they go to form the one percent which makes the whole possible."[9]

Fuller originally wrote these words in 1919, but that was after most of the damage was already done. In any case, we will not go as far as agreeing to the ninety-nine percent statement. Even with large superiority of weapons, engagements were lost because of poor leadership or sheer stupidity of underlings (such as at Isandhlwana; see chapter 3), but the facts of real life apparently had some effect, albeit late.

Where did the classical writers on military theory, from the seventeenth century and on, go astray? The importance of weapons as such to fighting ability was already treated in the Bible, and it appears that the ancients had a better understanding of the problems involved than the "modern" writers. In I Samuel 13:19, the Philistines forbade the Israelites to have blacksmiths, so that they could not manufacture swords and lances. This kind of thinking is extraordinary in its sophisticated approach (for that time) to the question of arms control—or more accurately to the ancient equivalent of "gun control"—even if it is doubtful if a rural militia force could do much with such sophisticated weapons if they had them.[10] But the important point is that the more recent authors missed completely a point that was obvious, and acceptable, to the ancients.

With few exceptions, weapon development was generally not carried out by the military, and rarely so to answer a specific military need. Weapons were occasionally thought about or improved by practicing officers, but most of the development was carried out by the artisans who built them or the amateurs who thought up better ways of doing things.[11] Furthermore, the one group of people who probably could have contributed the most to the quality of the discussion was missing. The real thinkers, the natural philosophers, were not involved in weapon improvement. The scientists were almost never exposed to the "right" environment. Consequently, they either did not have the right tools to deal with these problems or were not really interested. It should also be pointed out that until the French Revolution, men of science disdained occupying themselves with military problems as such. Most of them were good churchgoing Christians and took the command "Thou shalt not murder" quite literally. The few who did briefly touch on such matters, for example, Leibniz, Bernoulli, Huygens, and Newton, for example, used artillery firing to test their physical and mathematical theories. They would have been horrified at the suggestion that they were doing military R&D (research and development) and actually contributing to the development of guns. Leonardo da Vinci (1452–1519), who was deeply involved in "defense matters," was rather an exception for his times.[12]

Like many other conventions shattered by the French Revolution, so was the idea that for some reason men of learning are banned from participating in the military struggle of their countries. During the revolution and afterward, struggling revolutionary forces exhorted the men of science (and in fact everybody else) to contribute inventions and scientific advances to the cause, which essentially broke the taboo. Science was finally mobilized to improve the quality of the implements of war.

It is not enough to invent or develop a better weapon, however; it is also necessary to convince the military to adopt it. That has typically required overcoming deeply held beliefs and attitudes built up over centuries. Furthermore, as we have seen from Clausewitz's writing—and he probably was only expressing a generally accepted opinion—weapons were considered of secondary importance to other factors. Here we probably can find the roots of the mystique of the "attack," so prevalent in the European armies of the nineteenth and early twentieth centuries.

Sun Tzu, writing some twenty-five hundred years ago, may be excused on the grounds that the military technology of his time was rather primitive, or that he knowingly chose to avoid discussing hardware because of the cultural standards of his society and concentrated instead on the abstract. But the attitude of Clausewitz and his contemporaries about this subject poses some problems. After all, weaponry in Clausewitz's time had become quite sophisticated and varied, and he should have been aware of these various developments in weapons technology. These included widely used and varied (land and sea) artillery, many forms of handheld and shoulder-operated firearms, mechanical signaling devices, sophisticated ships of war and naval production and organization, mass production of weapons, and even a hint of aerial observation by means of balloons. Surely Clausewitz, who no doubt studied the classics and for all his career was a staff officer, knew and understood the differences between the equipment of the Roman cohort and that of the Prussian infantry square, and how one evolved into the other.

SOME POSSIBLE EXPLANATIONS

Initial achievements of the barefooted, half-clad armies of the revolution, admittedly supported by large numbers of conscripts, were fueled by enthusiasm. Furthermore, Napoleon, with no real technical advantage but with his operational genius, achieved stunning victories. The willingness of the troops and the ability of the commanders may have further influenced the thinking of Clausewitz and later writers.

Also, one may argue that senior officers of Clausewitz's rank did not stoop to consider mere tools of the trade as worthy of their attention, but this argument barely holds water. While some of them could be considered mavericks, there were some figures of note who did pay attention to this problem. Among many others, great national and military leaders like Peter I (the Great) Tsar of Russia (1672–1725), Gustavus Adolphus the Swedish king (1594–1632), and more recently the Frenchman Lazar Carnot (1753–1823) played a part in the actual development of weapons and their effective employment. Gustavus Adolphus, a renowned leader and organizer, introduced a highly mobile light artillery piece, which unlike heavy artillery accompanied the rest of the army to the battlefield. He also introduced as standard issue the unitary charge, which contained the powder and the ball in one paper cartridge, an invention from some eighty years earlier. Peter the Great traveled abroad and worked as a simple dockyard hand in a Dutch shipbuilding yard in order to learn the trade firsthand. He was later instrumental in creating, nearly from scratch, a Russian navy. He also contributed to the design of better muskets and artillery. Carnot, while not an inventor, organized France's scientific resources in support of its military. It is noteworthy that these three were all active before Clausewitz started his books, and presumably he knew of their achievements.

Another possible explanation is that, generally speaking, at that time innovations in weaponry were very slow and could not be made to order, even if one knew what or where to order. The concept of innovation and improvement in weapons as an orderly and desired process was barely understood, if at all. Thus the classical qualitative attributes of the warrior and his leaders seemed more important and worthy of greater attention. Military tradition simply could not accept the role of technology in improving battlefield performance, and Napoleon for one seemed to be living proof that "military genius" was enough. This attitude of course did not originate with Clausewitz. It was quite prevalent throughout history. In many cases it bred contempt for the enemy's weapons, slowed local improvements in one's own weapons, and was one of the main factors occasionally leading to technological failure. Alexander Forsyth invented his firing cap in 1805. In 1807, in the midst of the Napoleonic Wars he offered it to the British army, who promptly rejected it. It was finally adopted only in 1834.

It should be pointed out that with few exceptions, military officers were never against improvements in their weapons. They would always accept a longer-range rifle or a lighter radio. They might argue about the price, but they were often easy to convince. What they usually did object to were radical inventions or extreme improvements that necessitated conceptual changes (and changes in doctrine), required special support equipment, or came at the expense of old and tried, nostalgically loved, systems. Two examples will serve to illustrate the point. In the beginning of the nineteenth century, after failing to convince the French that he could provide them with submarines, steamboats, and underwater weapons, Robert Fulton tried to sell his "torpedoes" to the British navy.[13] He almost succeeded, but then on October 21, 1805, Admiral Horatio Nelson won the Battle of Trafalgar. Reasonably enough, the British lost their fear of French invasion, and the Earl of St. Vincent told Fulton that "Pitt was the greatest fool that ever existed, to encourage a mode of war which they who commanded the sea did not want, and which, if successful would deprive them of it."[14] Jervis St. Vincent (an admiral who was once Nelson's commanding officer) was an extremely capable and successful naval leader and recognized as such. Coming from such an authority, this verdict sounded the death knell for Fulton's ideas, at least for the time being.[15] Fulton returned to his native United States and, after some more work on underwater explosive devices, turned his attention to steam propulsion for ships, which his work helped to promote. Steam propulsion for men-of-war finally caught on in Europe, and the French, still smarting from their naval defeats, were willing to listen. On the other hand, the British navy, although accepting steam propulsion for auxiliary craft, did not want to adopt it for warships. Not to be outdone by his illustrious predecessor, Lord Melville, first lord of the Admiralty, answering a request from the

Colonial Office for a steam packet, wrote in 1828: "Their Lordships feel it their bounden duty to discourage to the utmost of their ability the employment of steam vessels, as they consider that the introduction of steam is calculated to strike a fatal blow at the naval supremacy of the empire."[16]

As will be discussed later, such denials of technological advances were not isolated cases in Britain or the rest of the world. Only when there was a combination of past disasters or long-lasting incompetence, which highlighted the need for reforms, and a capable leader with enough clout, were advances also made in the field of weapon innovation, and improved technology and organization were called upon to solve the problems.

THE PROBLEMS OF EXPANDING TECHNOLOGY

In the nineteenth century, many began to realize that industrial and technological innovation was shaping war. The introduction of the telegraph, the railroad, the steam engine as an enabler of massive production capability, metallurgy, chemistry, and all the myriad other resultant inventions, most of which were developed for civilian use, had a tremendous impact on military operations. This fact was understood even by the cavalry officers, and most of these men realized that they could neglect such developments only at their peril. However, the train and the telegraph were means to an end, since battles are not really fought with a train or a telegraph. So the military officers learned to use these new technologies to their benefit, and some of them, particularly in the more technical branches, even understood the scientific principles underlying these innovations.

This trend toward the increasing influence of science and technology on warfare gathered momentum and resulted in the First World War being a clash between the industries of the adversaries. Science did play a role, especially in chemistry, which had to supply substitutes for what could not be obtained from abroad because of unavailability or blockade, or by finding better ways to produce certain chemicals in huge quantities.[17] Toward the end of the war, physics joined the fray with pioneering work on antisubmarine detection systems and improvements in radiotelegraphy. On the other hand, the development of the tank or the improvements to the airplane, important as they were to advances in the science of warfare, did not depend on new scientific principles nor even on really advanced technologies. In the case of the airplane (during the First World War), it was more a matter of time to effect steady improvements.

However, officers bent on advancement shunned too close an association with the technical branches. For instance, in June 1920 the tank corps was abolished in the U.S. Army, and the tanks became part of the infantry. Still, there were officers who thought that tanks should be more than a mere auxiliary to the infantry. Later in 1920 Captain Dwight D. Eisenhower, who served in the armor branch (in the infantry), published a controversial article about tank warfare in the *Infantry Journal*. He was brought before the chief of the infantry, threatened with a court-martial, and chose to transfer out of armor. George Patton, too, reassessed his career plans as a tanker in the infantry and transferred back to the horse cavalry. There "he could play polo, participate in horse shows, and hunt."[18] Nor were the British, the inventors of the tank, free of guilt. "Finally, the inherent conservatism of the British army, centered on the Regiment and tradition, created formidable barriers to innovation, and these institutional realities were exacerbated by a general anti-intellectual bent that pervaded the Army's culture."[19] Even stronger language was used by another writer: "Within the Army the Technical Officer was generally despised."[20] Even in the British tank corps, many officers considered technology to be beneath their dignity.[21] This attitude was not limited to the British army, and we find traces of it also in the Royal Air Force (RAF). Frank Whittle, the inventor of the jet engine, was originally trained as a pilot in the RAF and also served as a test pilot and flight instructor. When posted by the Air Ministry to full-time work on his own invention, he was worried about the effect that this posting, to the so-called Special Duty List, would adversely affect his RAF career.[22] And this was at the end of 1937.

Between the two world wars, science and technology took giant strides forward and enabled the introduction of new equipment and new war-fighting concepts. These included a much-expanded role for aviation, the increasing mechanization of the ground army, and widespread introduction of wireless telegraphy (which by itself enabled the implementation of a new concept, "highly mobile warfare"). It also included aircraft carriers and the terribly complicated technology of landing aircraft on a tiny deck, which one pilot described as "something the size of an elongated postage stamp." These new technologies brought strategic bombing, a new kind of surface naval warfare with an emphasis on carriers, some improvements in torpedo designs, and the ability to operate over vast distance at a much greater speed. Both military and political leaders in most countries realized these changes and their effect on future warfare and made an effort to harness the fruits of the advances in science and technology to the conduct of warfare. In certain countries these efforts became more urgent as war approached. Not all countries, of course, handled these matters with the same degree of efficiency, nor did all of them have the capability to do so.

OLD KNOWLEDGE VERSUS NEW, AND THE EFFECTS OF SPECIALIZATION

But there were several problems lurking beneath the surface. First of all, scientists, military officers, and political leaders are humans, with all the problems that a diversified group of capable (at least in their own specialized fields) human beings can generate, even within one field of occupation. Second, the fields of science, and the various resulting technologies, and the myriad possible products based on these technologies were expanding at an ever-accelerating rate, both in the number of areas they affected and in the depth of this effect. This led to specialization. Well into the nineteenth century, a "scientist," even an engineer, was expected to know almost everything about his profession.[23] This was not because he was expected to be an extremely gifted individual, but because the sum total of available knowledge was such that a fairly skilled person could encompass it all. This circumstance changed in the twentieth century, a little slower for the pure scientist but fairly quickly for the engineer. The trouble for both was that old knowledge could not be completely discarded, so in order to stay competent the professionals had to specialize. This in turn led to another problem. Practical problems are often multifaceted, and professionals focusing on a particular problem often did not realize that occasionally another scientific discipline had the solution (or just that little nudge in the right direction) for which they were searching. The problem was eventually solved when the concept of "interdisciplinary effort" was formulated and accepted as useful. In the meantime, many scientists were working in isolation, occasionally trying to solve by themselves problems that were trivial to scientists in other fields, and at best wasting time and effort.

If this situation was slowly getting bad in the scientific community, it was bad from the start in the military and even worse with political functionaries. Military professionals understand their profession, or are supposed to, although as will be demonstrated, on occasion this is far from the case. Be that as it may, during their careers they have no time, and even less inclination, to become proficient in technical subjects except for very specialized cases like navigation for naval officers, or trigonometry for artillery people, or some electronics and communications theory for the signal branches. An infantry or cavalry officer does not need even this much science. The future high command of any military force comes from the combat arms and not from the support services. In the ground forces high command came mostly from the infantry or cavalry. These people were starting to be confronted by problems that even the scientists were occasionally in dispute about, and they quickly lost interest, if they ever were interested in them at all. Science simply became too complicated. Future potential developments bordered on science fiction, and few serving officers were willing to take them seriously.

Scientists on occasion also did not endear themselves to military personnel. Civilian men of science always saw themselves as above the petty squabbles of humankind. The Brodies relate a story about Ernest Rutherford, the noted British physicist, who was once chided for being absent from some committee meeting about submarine warfare (of which he was a member) and supposedly answered: "Speak softly please. I have been engaged in experiments which suggest that the atom can be artificially disintegrated. If this is true, it is of far greater importance than a war."[24]

And finally to the politicians. With one noted exception, that being Japan until 1945, military institutions in all kinds of regimes depend on the politicians to dole out money for the maintenance and development of the armed forces, including the salaries of the generals. But politics is probably the only profession where no test of competency is required in order to practice it. What's more, some regimes were (and often still are) run with the mores and customs of a pirate ship. Even in the real democracies, the acquisition of a senior-level management position occasionally depends more on loyalty to the cause (or on the size of the campaign contribution) than on real competence. There is some logic to this. If the man is rich enough to contribute generously, presumably he is smart enough to begin with.

THE PROBLEMS OF COOPERATION

When these three groups of people—the scientists, the generals, and the politicians—had to work together for a common cause, it was only natural that there would be friction, and one of the first concepts every engineering student learns is that friction creates a lot of useless heat. This sort of attitude was a perennial fact of life and persisted well into the Second World War. For example, the difficulties between the scientists of the "Manhattan Project" and General Leslie Groves, who headed it, are related in every history of the development of the atomic bomb. In another case, the priorities for waging the war were set in conferences between the Allies. A policy of "Germany first" was agreed upon, and within this framework it was decided that winning the Battle of the Atlantic was of the utmost importance. If this battle was not won decisively, the buildup of materiel in Britain, for the return to the continent and for the waging of the strategic bombing campaign, would be hampered. It was that simple. However, many influential officials (see chapter 5) considered the bombing of Germany more important and refused to allocate the necessary aircraft to antisubmarine work.[25] The generals usually said that they would be damned if they would let longhaired civilians teach them their job. The admirals were no better. Admiral Ernest J. King, chief of naval operations of the U.S. Navy during World War II, displayed just this kind of attitude. King refused to understand the contribution of several new developments in the radar field (developed by civilian agencies) to a revised approach to antisubmarine warfare. When the design of a new cruiser was presented to him, he complained, "There's too much radar on this ship. We've got to be able to fight a ship with or without radar."[26] Furthermore, King is described as having questioned the American political system that permitted civilian intrusion into military affairs.[27] On another occasion he told Vannevar Bush, the head of OSRD (Office of Scientific Research and Development, at the time one of the more powerful agencies in Washington), that "civilian opinions didn't interest him."[28]

Another problem was interservice squabbles over priorities in receiving new equipment and over roles and missions. The case of Billy Mitchell is typical. When in the early twenties he proved that aircraft could actually sink ships, the army, of which the army air corps was a part, demanded a piece of the naval action. After some heated arguments, a compromise was reached by which the army air corps was to defend American shores to a distance of one hundred miles, and beyond that it would be the navy's responsibility. However, the army air corps, who supposedly prepared themselves to guard the shores against enemy battleships, never trained their pilots to fly over water and thus failed when German submarines started their depredations along the eastern seaboard (the navy was blamed for that failure). Another problem occurred in Britain when the new ten-centimeter (wavelength) radar sets were being introduced and Fighter Command, Bomber Command, and Coastal Command were all vying for them with complete disregard for the overall picture. The situation on the German side was hardly any better. The Luftwaffe had to be ordered to contribute aircraft to support the German navy, and Goering fought (unsuccessfully) against the navy's efforts to develop a torpedo to be carried by aircraft.[29]

There was also the problem of jockeying for political power and influence, and the meddling of high-ranking politicians in technical and operational problems they did not understand, or applying long-past and irrelevant personal experiences to completely new technical and operational situations. The German fixation with dive-bombing is a case in point. Ernst Udet was the second-highest-ranking German fighter ace of the First World War, and he was a firm believer in dive-bombing. Between the wars, he performed as a stunt pilot and in 1936 was persuaded by his old comrade Hermann Goering (also an ex–fighter pilot) to return to the Luftwaffe. His first position was as inspector of fighters and later he was promoted to the position of quartermaster general of the Luftwaffe. In this post he used his influence to ensure that no airplane was accepted into the Luftwaffe service unless it could operate as a dive-bomber. While the light bombers could be coaxed to dive—and more important, to come out of the dive—it was a completely different matter with the heavier aircraft. Even the four-engine (but driving only two propellers) He-177, intended for long-range maritime patrol, was required to show this capability. Since the He-177 was manifestly incapable of this kind of performance, Udet lost interest and the airplane lingered on, plagued by technical problems (owing to its unusual engine arrangement) and without any real effort to resolve its shortcomings and bring it into service. Finally, there were the usual small and large inefficiencies extant in every large-scale bureaucracy such as the armed forces. Under the circumstances it is quite surprising that that much progress was achieved by all involved, but as we shall see, there were also a lot of failures.

THE EFFECTS OF LARGE AND SMALL FAILURES

While in older times a defeat in a single battle decided the outcome of a campaign or even a war, a modern war is usually too involved to be decided by one battle, or even a series of them. But occasionally there are certain climaxes that with twenty-twenty hindsight we can single out as turning points. The outcome of such battles or campaigns either decided the outcome of the war or robbed one side of a sure and relatively quick victory. The Battle of Britain and the submarine campaign in the Atlantic are two examples of such German failures. On the Japanese side there was the Battle of Midway.

There were also many small failures and lost opportunities that, although none of them alone was big enough to be irreversible, their accumulated effect in casualties and drain on resources did become critical. During a long war, such little failures could start a snowball effect, the results of which are difficult to predict. Considering the harsh competition for resources during an actual war and the pressure of current crises, rectifying past mistakes under such conditions is next to impossible.

After the dust settles, the victor usually sweeps these little failures under the rug with the comment that this is the price of victory. In the eyes of the defeated, such failures usually grow in size and stature, occasionally leading to a great deal of soul-searching and eventually real improvements. The true proportions and relative importance of these small failures come to light only slowly, as personal observations become public and a new generation of historians exposes the full range of facts. The losses, however, stay lost, and there always remains the nagging question of whether victory could have been achieved at a lower cost.

LACK OF PREPAREDNESS

One form of technological failure can occur when one side is well aware of the implications of a new weapon system in the hands of the opponent, but for various reasons does nothing about it, at least for the present, because of political illusions or more pressing priorities. One such instance was the British lack of preparedness for World War II until it was almost too late, and the frantic efforts to close the gaps when the danger was realized.

Another example was the case of the Soviet Union and the atomic bomb, although here the political leadership could be blamed only remotely. The Soviet leadership was quite well informed (through its extensive spy network) of the progress of the United States toward an atomic weapon. However, although they had a modest nuclear effort of their own, which dated back to the thirties, except for gathering all the theoretical and practical data possible, they were not in a position to do anything about it during the Second World War. After the United States ended the war with Japan by the use of atomic weapons, the Soviet leadership realized the significance of this new technology in world politics and made huge efforts to master it too.

The Soviet Union exploded its first atomic device on August 29, 1949. At that time the United States already had some 170 nuclear bombs. By 1951 the Soviet Union had 25 and the Americans had 438. These numbers grew in 1953 to 120 and 1,169, respectively. The Soviet Union was well aware of this discrepancy in numbers and of its inferiority in this respect and did its utmost to overcome the problem, but in the meantime all it could do was sit tight and hope. Admittedly this did not stop the Russians from occasionally being pushy (such as blockading Berlin in 1948), but they were most careful not to step over the line. This late arrival on the scene had another interesting by-product. Although in the thirties the Soviet Union led the world in the development of very large aircraft, including bombers, this field was totally neglected during World War II. The Soviets had to concentrate on air superiority and on aircraft for support of ground warfare, the most famous of which was the IL-2 "Stormovik."[30] When the war ended, the Russians were without a real long-range bomber and with no hope of producing one in a reasonable time to carry atomic weapons. They did attempt some temporary stopgap measures such as hand building the Tu-4 "Bull," a copy of the B-29. Several of these aircraft landed in Siberia when they ran into trouble on bombing missions to Japan. The Russians returned the crews but kept the aircraft. The Soviets went on to develop some bomber capability but decided to jump-start their strategic buildup by going directly to rocket-powered missiles. Although the Soviets had a long tradition of rocket research and an impressive expertise in solid fuel rocketry, they chose to copy and improve the German V-2 design.[31] This approach paid handsome dividends, in both the military and the scientific fields. These powerful rockets enabled the Soviet Union to obtain some important firsts in the space race, and it held the lead until the United States managed to catch up. A similar, almost reciprocal case was the so-called Missile Gap. Based on the Soviets' demonstration of their prowess in the rocket field, the American leadership assumed the worst and made a huge effort to close this gap. In the meantime, the American public lived in fear of an impending Soviet attack, which they believed could not be thwarted or even avenged.

While there were more such hypothetical cases of "what would have happened if," another case—a more concrete one—was the problem that the British, and later the Allies, had with the German U-boats in the Atlantic Ocean during World War II. During World War I, the submarine proved an unanticipated menace and possibly came closer to strangling England than it ever did in World War II. In mid-1917, for example, the British battle fleet was confined to port because of lack of fuel, and losses of shipping in April 1917 amounted to some 860,000 tons.[32] In comparison, the high-water mark in World War II shipping losses came to 601,000 and 627,000 in May and June 1942, respectively, and 538,000 in March 1943.[33] In 1917 the British finally instituted the convoy system, which previously had been rejected twice for completely untenable reasons, and losses dropped dramatically. It appears that for all their knowledge of trigonometry, many leading admirals had problems with plane geometry.[34] The system of convoys had originally been used, quite successfully, against French commerce raiders during the Napoleonic Wars. Also, toward the end of the war the British developed the ASDIC (later known as "Sonar"), but it was not ready when the war ended.[35] The British remembered the lesson and in the thirties pushed the development of an improved version of ASDIC, somehow managing to keep this development a secret. So at the beginning of the Second World War, they were pretty confident of their ability to stop the German U-boats. Admittedly, some German submarine commanders received a nasty surprise, but this did not last long as the Germans started attacking at night on the surface, where the ASDIC was inefficient. The British Admiralty, which knew about this shortcoming of the ASDIC, promoted its self-confidence by claiming that submarines would not operate at night on the surface because it was too dangerous. This ruling was based on a safety measure, instituted by the Admiralty for British submarines and did not really apply to the German navy.[36] Another problem with the early ASDIC was its range, which was much shorter than the torpedo's range or the range of the listening devices on the submarines. Convoying also helped, but not enough, and losses mounted alarmingly. For the Allies it was an uphill struggle to develop, produce, introduce, and integrate all the means that the scientists dreamed up, and worse, it took time. In the meantime ships were being lost, but there was no magic wand to speed things up. Finally, in mid-1943, all the effort came together and the submarine in essence was defeated.

HOW DO TECHNOLOGICAL FAILURES OCCUR?

All of these big and small successes and failures depended on many factors, from the amount of resources, and their quality, to the quality of the leadership. But once in a while the outcome depended on some technological fact, understood correctly and used, or for some reason misunderstood, disdained, neglected, mismanaged, or meddled in by the politicians. The following chapters will study several examples for the various reasons that technological failures did occur. These reasons may be divided into the following five groups:

1. Conservative thinking, mistrust of new ideas, and inability to adapt to changing environment
2. Misunderstanding of the technology involved or its relevance to the battlefield
3. Poor management (including the NIH—Not Invented Here—attitude) and bad leadership

4. Preconceived notions by very important persons, sometimes accompanied by overconfidence and arrogance
5. Meddling by higher authority, sometimes because of political ideology

This list is not graded according to importance of the different causes, the amount of damage they did cause, or the frequency of documented cases. Also, some of the failures described can have more than one cause, and thus the list should be regarded only as a general guide.

If we accept the basic premise that there were cases when an available innovation was not introduced or appreciated, then the inescapable conclusion is that people, and not laws of nature, are to blame. Going even one step further, there are cases when the scientists or engineers are the cause of delayed appreciation of technological innovation, but on the whole this is less common.[37] Usually, scientifically trained people appreciate fairly well what is possible and what is not. They also are less apt to be swayed by slogans or nostalgia. The problems start when judgment and the power of decision are in the hands of those who were trained in other disciplines, but who exercise control over matters of technology. Admittedly, this is a generalization, and technical people also make mistakes, but generally their batting average, as far as future technology is concerned, is higher.

Some of these failures were caused by unquestionably patriotic figures with the best of intentions, but who were misguided by previous experience or past traumas. Last, these reasons for failures were not, and are not, limited to any particular historic period, country, social stratum, culture, or form of government. Fools, skeptics, and arrogant decision makers have come from the ranks of kings, rebels, communists, Nazis, and true democrats. One thing is certain: it will remain an everlasting problem. To paraphrase Kipling's famous saying, such stumbling blocks to military technological progress may fade away or die, but they are always replaced by new ones.

The following chapters will deal with several groups of case studies involving novel military technologies and in the widest sense the treatment of military technological problems. In many cases individuals, usually in high positions, could, and often did cause enormous damage because of vanity, misplaced motives, arrogance, or sheer stupidity. Many of the problems described could have been solved in days or weeks but occasionally took months, and even years, to straighten out. Since most of these happened during actual wars, one can envision the waste and the unnecessary cost.

SOME OPINIONS OF MILITARY LEADERS

However, before going into the more detailed case studies, and to set the tone, the following are high-ranking military officers who in one way or another did not fully appreciate emerging technological innovations, all of which later proved quite successful.

As early as 1910, Helmuth Moltke (the Younger), chief of staff of the German army at the beginning of World War I, thought that the use of the airplane to drop bombs was "for the present unimportant." Nor did War Minister General Erich von Falkenhayn consider, in January 1914, the airplane of much use in a long war.[38]

Ferdinand Foch, during an aviation exhibition in 1910, commented on aircraft: "That's good sport, but for the army the aeroplane is of no value."[39] At the time (1907–1911) he was commandant of the French École Supérieure de Guerre, influencing generations of future senior officers.

In 1799 Napoleon disbanded the two balloon companies then functioning in the French army, viewing them as useless.[40] Furthermore, he could not make up his mind with regard to Robert Fulton's ideas about underwater explosives, submarines, and steamboats. When he did decide that the steamboat had potential, it was too late; Fulton had left for England.[41]

General George Marshall and General H. H. Arnold, in the United States in 1942, did not consider a new antisubmarine radar important enough for them to observe a demonstration. They finally did so after being ordered to by Henry Stimson, the secretary of war.[42]

Ernst Udet, in 1938 and at the time quartermaster general of the Luftwaffe, when shown a ground-control radar, exclaimed "Good God! If you introduce that thing you'll take all the fun out of flying." The man who showed him the radar commented that "he [Udet] was, of course, an old aviator of the First World War, and I rather think he still lived in those days."[43]

3 Misunderstanding the Battlefield Environment

Fools say they learn by experience. I prefer to profit by other people's experience.

Otto von Bismarck

MACHINE GUNS IN THE FIRST WORLD WAR

Introduction

For centuries inventors and military men tried to increase the rate of fire and the reliability of handheld firearms. The powder ignition system went from the slow-burning match, to the wheel lock, and on to the flintlock and percussion cap. But as long as the guns were muzzle-loaded, the loading process itself was cumbersome and slow. Some increase in firepower was achieved by bunching a number of barrels and firing them together. While this had the effect of a powerful simultaneous volley, which had its uses on the battlefield, the reloading process remained lengthy and the overall rate of fire over time did not improve significantly.

The introduction of the percussion cap, together with the brass cartridge, finally enabled the development of a reliable breech-loaded gun. This in turn enabled the development of the first mechanized loading, firing, and extraction system. This was the Gatling gun, so named after its inventor, Richard J. Gatling, a medical doctor who was more interested in developing mechanical contrivances than in pursuing his medical career. Gatling in fact developed several agricultural machines, particularly various seed spreaders, that had a considerable effect on agriculture in the

United States. Some of the ideas from these machines later served in his machine gun. Purists claim that the Gatling was not a truly automatic system, because the motive power was supplied by the gunner turning a crank. Nevertheless, this enabled a tremendous improvement in the rate of fire. It could fire at the rate of about two hundred rounds per minute, depending to some extent on how rapidly the crank was rotated. The Gatling was invented in 1862 but perfected toward the end of the American Civil War, and was slowly adopted by the military forces in many countries. However, the weight of the system, with its multiple barrels, was considerable. It was mounted on a two-wheeled carriage like a field artillery piece, which occasionally led to poor tactical use because it was treated like an artillery piece, which it was not. Finally, its weight made it difficult, although not impossible, to advance with the infantry in an attack.

Early Developments

Hiram Maxim was a successful and wealthy American inventor, who pursued his interests in steam machinery, chemistry, electricity, and aviation. Historian John Ellis relates how on a visit to Vienna (to a technological exhibition) in 1882, Maxim met an old American acquaintance who told him, "Hang your chemistry and electricity! If you want to make a pile of money, invent something that will enable these Europeans to cut each others' throats with greater facility."[1] Maxim took this advice to heart, moved to London, and started working on designing a better machine gun than the Gatling. Maxim was the first to use the energy of the recoil of the gun when fired to extract the empty cartridge, then load a new one and fire it.

In 1884, Maxim perfected his gun and demonstrated it for several leading figures in England. The new invention was initially accepted enthusiastically and there was no question of its efficiency or reliability. Maxim ran into difficulties on the commercial side so he teamed up with Thorsten Nordenfelt of Sweden, also a manufacturer of "machine guns." The Nordenfelt weapon was actually another crank-operated gun, similar to the Gatling, but

Nordenfelt had a better salesman, and apparently the value of better public relations in selling a product, no matter what its quality, was raising its ugly head even then. The new partnership succeeded in selling the Maxim guns, but not at a rate that made them rich. Some military forces bought the guns and some were sold to colonial governments and to various expeditions going to Africa, where they were used extensively and successfully in the suppression of native rebellions. The market became even more crowded in 1892 when John Browning developed his gun, and soon afterward the French Hotchkiss and others became available. But during the early years, the machine gun became synonymous with the Maxim name.

Opposition to the Machine Gun

Strangely enough, the established military forces were not very enthusiastic. Many reasons were given to explain this attitude, most of them centering on the cultural outlook of the officer class of Europe's armed forces. In brief, the reasons explaining the reluctance of the European armies to adopt machine guns were as follows:

1. The mechanization of warfare, exemplified by the machine gun, would rob the soldiery of its traditional values of skill, valor, and honor, in essence making man subservient to machine.
2. The traditional officer corps in all countries did not understand technology and thus hated it. They assumed that if technology were ignored, it would simply go away.
3. The military establishments did not realize that the battlefield was changing at an accelerated rate. This was due to the introduction of previous innovations in firearms, such as the magazine-fed, bolt-operated rifles; quick-firing artillery; and explosive shells.
4. The culture of the attack was the predominant doctrine of the European armies. Obviously, charging cavalry did not have much use for the machine gun. Attacking infantry, on

the other hand, would be slowed by the need to manhandle these heavy contraptions.

5. The last factor would tend to promote a defensive mentality, an anathema to traditional thinking.
6. Finally, machine guns were expensive and they consumed ammunition in prodigious quantities. Purchasing sufficient numbers of machine guns would be extremely costly.

Early "Combat" Use of Machine Guns

Each of these arguments contained a grain of truth here and there. So they seemed reasonable at the time to the high-ranking officers (and nobody really consulted the junior ones), but the worst blunder made by the officers was that they totally refused to adapt to the facts of real life.[2] As pointed out earlier, colonial forces of the very same European armies had used machine guns with great effect in Africa. The British used Gatlings during the second expedition against the Zulus in July 1879.[3] In 1885, a Gardner machine gun saved a small detachment of British soldiers from a mass attack by the Dervishes, during an attempt to relieve Major General Charles Gordon in Khartoum in the Sudan. Gatlings also saw action in the Battle of Tel-el-Kebir in 1889 to suppress an Egyptian threat to the Suez Canal. In 1898, Maxim machine guns provided the margin of victory in the Battle of Omdurman in the Sudan.[4] On numerous occasions in between, government troops and various trading and commerce enterprises that moved into black Africa used Maxims to extend British rule there. Finally, machine guns were extensively used in Africa during the second Boer War (1899–1902) and as late as 1906 in one of the last rebellions of the Zulus.

Not to be outdone by the British, in 1890, the German East African Company in Tanganyika tried to copy some of the British practices. It, too, ran into natives who did not properly appreciate the "white man's burden" and resolved such difficulties in the same manner; that is, by force of arms, greatly enhanced by machine guns. In 1891 the German government supplanted the East African Company, but it, too, quickly had a rebellion on its hands. Only by 1905 was this uprising quelled, and again by sheer firepower. In these engagements it was proved conclusively—and the Europeans accepted it as a tactical constant—that a few well-placed machine guns could stop an attacking force, almost without regard to its size, from reaching the defensive line. A popular ditty at the time ran: "Whatever happens, we have got / The Maxim Gun, and they have not."[5]

In addition to Africa, British forces in Burma and in India on the North-West Frontier used machine guns. During the Boxer Rebellion in China, in 1900, the defenders of the European legations besieged in Peking used machine guns, as did the relief column that came to their aid. British forces used them in a punitive expedition to Tibet in 1904.

The Russo-Japanese War

In 1895 Japan defeated China in a brief war. The spoils of victory included the strategically important Liaotung Peninsula and harbor at Port Arthur; however, France, Germany, and Russia forced Japan to relinquish Port Arthur, which, once surrendered to Russia, became a major naval base. In February 1904, during a clash of interests with Russia over Manchuria, the Japanese decided to try to reconquer Port Arthur, which they deemed was theirs by right of conquest. Without a formal declaration of war, they opened hostilities with a surprise night torpedo boat attack against the Russian fleet at Port Arthur, causing damage to several ships. Subsequent attacks penned the rest of the Russian fleet in the harbor, in part by sinking several merchant ships at its entrance. Beginning June 1, Japanese ground forces laid siege to the city. Because this was Russia's only ice-free harbor on the Pacific Ocean, the Russians defended it stubbornly.

For a European army, and in contrast to all other such armies, the Russians were great believers in the power of the machine gun. They tested the first Gatlings in 1865, started purchasing them in 1868, and subsequently negotiated a license to produce them themselves. They used these weapons extensively in the war with Turkey (1877–1878) and in various cavalry actions in central Asia. They also considered machine guns an integral part of the infantry. They became so enamored with their Gatlings that when Maxim first displayed his wares, they were skeptical about guns without cranks. In the end, however, they were willing to be convinced and again wound up with a production license. The Japanese lagged behind the Russians early on, but they learned quickly and purchased French Hotchkiss weapons.

With two well-equipped forces squared off against each other in a modern war, all the major powers sent official observers and war correspondents to the scene. A lot of useful information was free for gathering. The Japanese army, though quite modern in its equipment, was somewhat old-fashioned in its tactics and conducted many frontal infantry attacks against well-prepared positions. Admittedly, they did not have many options, given the terrain and the placement of the fortress. Still, these methods had already proved very costly in the American Civil War and were definitely considered such in the recent Boer War, in which machine guns had been present but combat still dominated by rifles.

The Japanese, however, persisted in this mode of warfare. During one week in December 1904 they suffered some eleven thousand casualties in repeated attempts, which finally bore fruit, to take one hill. The placement of heavy guns on that hill brought about the Russian surrender on January 2, 1905, so the Japanese commander may have considered this a worthwhile investment. Total Japanese combat casualties numbered nearly sixty thousand dead and wounded.

Machine Guns in Europe

All these facts were widely described in the European press and in official reports, and everybody commented about the effects of machine guns against infantry in the open. Some comments were directed against the Japanese disregard for human life. Nevertheless, for the British army in particular, the machine gun remained a despicable weapon, fit only for use against colonial uprisings. Commanders had no choice but to accept being equipped with them, but they had the power to do so only in small numbers, say two to a battalion, and then leave them in storage during maneuvers. At the time, this ratio was attributed to the opinion of Douglas Haig, initially a corps commander and later the commander of the British forces in France during World War I. As will soon be discussed, this contempt for the machine gun apparently clouded even his strategic thinking. This attitude of the high-ranking professionals inevitably percolated into all branches of government, and when the British army in 1907 made a halfhearted attempt to obtain money for more machine guns, the treasury refused. The high command quickly gave up, probably considering this refusal a blessing in disguise. However, the other two major powers, the Germans and the French, took a somewhat different route.

A member of the German royal family on a visit to England in 1887 was given a tour of the Maxim works. He was impressed and arranged for a gun and an instructor to be sent to Germany to introduce the weapon to the German army. Eventually the gun was exhibited to the kaiser, who was immediately won over by its potential. Because the kaiser's word carried considerably more weight in Germany than did the queen's in England (where members of the royal family tried, without success, to push the Maxim), the German army was rather quick to adapt to the new mode of warfare, at least as far as equipment was concerned. In 1908 a huge sum of money was allocated for machine gun purchase and a large effort was made to make the troops aware of its potential effectiveness on the battlefield. Conscious of these most recent German efforts, the French also became interested and, after some parliamentary haggling, started purchasing machine guns in quantity and equipping the troops.

While all the European armies did finally acquire machine guns—presumably to mow down the adversary if he would be foolish enough to attack—absolutely nobody gave any thought to the other side of the question: what would happen if that very same adversary looked at them through the sights of *his* machine guns? Part of the problem likely derived from the experience of European troops shooting down black rebels by the hundreds, occasionally by the thousands, and it was amplified by the reports from Port Arthur. To the European mind of that period it was quite obvious that the huge casualties suffered by nonwhites were due to their "inferiority" relative to the white machine gunners. Ethnic superiority complexes pervaded most of the European nations: the Boche, or Hun, were considered inferior to the French, the Froggies were considered inferior to the Germans, but both were considered greatly superior to the Limeys. Consequently, it is easy to see why the "culture of attack" held on even in the face of a major acquisition effort of machine guns by all concerned. "We will easily stop them from attacking, but when we go on the attack, our moral superiority or élan will carry the day" was the prevailing attitude in all the armies. This curious approach was bolstered by "mathematical" proofs of the superiority of the firepower of the attacking force, which nevertheless disregarded the cover and the steadier aim of the defender.[6]

World War I Casualties

When the war started, the Germans tried for a quick victory, but changes in the original Schlieffen Plan and lack of consideration for the realities of war effectively stopped them in their tracks. They apparently forgot one of the more important teachings of their own Count Helmuth von Moltke (the Elder, 1800–1891), who said that "no plan survives the first contact with the enemy." The war deteriorated into the infamous trench warfare, and the generals soon found out that neither élan nor moral superiority would overpower machine gun fire. Not that they did not carry the experiment far enough. Table 3-1 depicts some of the more famous battles of the western front and the numbers of casualties.

In Table 3-1, the term "casualties" refers to both killed and wounded. It should be remembered that the ratio of killed (including those who died shortly thereafter) to wounded was about one to two and a half; namely, that from about one-quarter to one-third of the casualties were actually killed. As a reference, the number of casualties in the Battle of Gettysburg, during the American Civil War, was about twenty thousand for each side, although admittedly the Civil War, on the whole, was a much smaller incident than the First World War. It should also be understood that

Table 3-1: Battle Casualties in Certain Offensives in the First World War[1]

Battle	*Dates[1]*	*Attacker*	*British*	*French*	*Germans*
Ypres I	14/10–11/11, 1914	Germans	58,000	82,000[2]	130,000
Artois II	9/5–18/6, 1915	French	—	100,000	75,000
Loos	25/9–8/10, 1915	Allies	60,000	190,000	178,000
Somme I	24/6–18/11, 1916	Allies	420,000[3]	195,000	650,000
Ypres III	21/7–6/11, 1917	British	400,000	—	65,000[4]
Somme II	21/3–4/4, 1918	Germans	160,000[5]		150,000

1. Dates are given by the European convention, day, month, year.
2. Including 32,000 Belgians.
3. Including 20,000 British killed in one day (July 1, 1916), the highest number of British killed during a single day in any war.
4. Including some 9,000 prisoners.
5. Combined British and French casualties. In addition, the Allies lost some 70,000 prisoners.

although the "attacker" in the third column denotes the side that initiated the offensive, this was not a clear-cut case of one side continuously on the offense and the other on the defense. Because of the length of the campaigns, or offensives, in practically all cases there were numerous counterattacks and certain positions repeatedly changed hands. To put the numbers in the proper perspective, we should remember that the total number of dead in the armed forces of all belligerents (British Empire, French Empire, Russia, the United States, Italy, Germany, Austria-Hungary, Turkey, and Japan) during the war came to about ten million. This number reflects all the campaigns over the four years of the war and includes naval warfare, air warfare, those who died from sickness not related to combat, and those who died in prison camps, especially in Russia.

On the western front alone the numbers were as follows:[8]

Table 3-2: Casualties on the Western Front in the First World War

Nation	*Killed*	*Wounded*
Germans & Austrians	1,500,000	3,500,000
French (including Belgians)	1,264,000	4,045,000[9]
British Empire	850,000	2,000,000

There is no doubt that the cynical advice (cited above) given to Maxim in Vienna was perfectly right.

As pointed out earlier, the long unwillingness of military commanders to accept the machine gun or to introduce it into the force structure could be explained away as a lack of understanding of its potential or as the rearguard action of the old guard, steeped in old-fashioned military traditions. But this was true only until after the first battles of World War I were fought. No such easy way out is available to explain the persistence of the high command to continue its futile attempts to overcome the firepower of an industrialized military by means of willpower and fighting spirit. This comment applies of course to the generals on both sides. Admittedly, both sides made attempts to overcome their difficulties on the western front—the Germans by introducing gas and the British by introducing tanks. When neither worked as expected, partially because of the shortsightedness of these very same generals, described later in this chapter, they went back to the old and tried ways of mass attacks.

Although it is easy to blame the generals, it is much more difficult to understand the willingness of the political leaders to accept such casualty figures without replacing those responsible for such debacles. It should be pointed out that during the American Civil War, senior generals were sacked out of hand for considerably less. While this thoughtless (and useless) aspect of the conduct of the war on the German side can be shrugged away by claiming that under the autocratic rule of the kaiser anything was possible, this argument does not apply to the Allies. Britain and France were led by supposedly democratic governments, which were subject to a popular vote. While admittedly the intervention of the political leadership in the conduct of military operations is to be avoided, such casualty rates were by themselves a political issue. Even more surprising is the fact that mutinies in the armed forces and unrest among the population erupted only toward the end of the war.

Problems of Firepower in the United States Army

The story of the machine gun and its introduction into the armed services cannot be concluded without mentioning the role that the

U.S. Army played in it, particularly the aversion of certain influential individuals to improvements in the firepower of soldiers. As may be apparent by now, the main developers of modern machine guns were all Americans; namely, Richard Gatling, Hiram Maxim, and John Browning (the Vickers was essentially a Maxim design). It should also be noted that the first effective multishot pistols and rifles (Colt and Spencer, respectively) were developed in the United States. One of the explanations offered for this trend was that the American continent, in its rapid development, was always short on manpower, and thus American ingenuity naturally turned to automation and other labor-saving devices and stratagems. Of course, the epitome of this approach was the invention of the assembly line.

It is therefore extremely odd that in general the "official establishment" of the U.S. Army repeatedly resisted the introduction of better or quicker-firing firearms for the soldier. The opponents of such improvements did not even have the high-minded reasons of their European colleagues in this matter.

One of the first such obstructionists, and probably the most notorious of the lot, was Colonel James W. Ripley, head of the Ordnance Department of the Union during the Civil War. Ripley came into this position in April 1861, just as the Civil War was starting, replacing Colonel Henry Knox Craig, an old (literally) veteran of the War of 1812. At sixty-seven (somewhat younger than Craig), and after forty-seven years of service, Ripley was brought in to put Ordnance Department affairs in order. A good organizer and logistician, who spent most of his career in staff positions, mostly in ordnance, Ripley knew next to nothing about tactics or the importance of the weapons' technical/tactical performance in the field. Admittedly, he did several useful and noteworthy things. Because he was a great believer in standardization and economy, he reduced the number of types of artillery ammunition from some four hundred to about one hundred and organized a well-oiled ordnance supply system to the troops in the field. However, this sense of frugality and economy was also one of the worst problems that faced (actually menaced) the Union soldiers.

Ripley was against breech-loading rifles, Gatling machine

guns, observation balloons, and all other "new-fangled" ideas. He maintained that there was no point in equipping the soldier with the more expensive breech-loader when he would do as good a job with the muzzle-loading, smooth-bore musket—and save ammunition to boot—by lowering the rate of fire. Furthermore, Ripley believed that a lower rate of fire would force the soldier to take a more careful aim, thus achieving better results.[10] In Ripley's defense, it can be said that this argument persisted until after the Second World War. Stanton, the secretary of war, understood the pertinent issues even less, but he supported Ripley in the face of the army generals, and even President Lincoln could not budge him from his position. During Ripley's tenure, the army acquired some seven hundred thousand muskets, but only eight thousand breech-loading rifles. When an effective system was developed to convert the muzzle-loaders to breech-loaders, Ripley dragged his feet until the proposal quietly died. The Henry rifle, with a magazine of fifteen rounds, was invented in 1860, presented to the Ordnance Department at the time, and rejected. It was presented again to Ripley and rejected again, on the grounds that the mechanism was too fragile. Only about seventeen hundred were bought by the Union army, but some ten thousand were bought privately by various regiments (who could afford quality), and it proved its worth in combat. The Henry was later developed into the Winchester repeater, which was considered "the gun that won the West." But the spirit of the Ordnance Department lived on. When in June 1876 General George Armstrong Custer made his famous last stand in the Battle of Little Big Horn, his soldiers were equipped with single-shot, breech-loading, .50-caliber Springfield carbines. These older rifles replaced the seven-shot, Spencer repeating rifles with which the cavalry had been armed until then, and which for some unfathomable reason were taken away from the troops several weeks before that particular campaign.[11] The Sioux and the Cheyenne under the leadership of Crazy Horse and Sitting Bull (and a less famous Indian chief named Gall) were armed with Winchester repeaters. As Clint Johnson put it, "The Indians' purchasing was not controlled by the Ordnance Department of the United States Army."[12]

As usual, the story is a bit more complicated. Ellis writes that Custer had four Gatlings with him, but on that fateful day declined to take them so as not to be slowed down, although "they were of a model specially designed to be dismantled and carried on pack mules."[13] Bruce Rosenberg also gives details of Custer's refusal to take more troops and his refusal to take Gatlings, although here only two Gatlings are mentioned.[14] This of course tallies with Custer's customary arrogance and his disdain for the Indians, epitomized by his reckless behavior on that day. It remains a moot point if a different rifle would have made a difference in that battle, although it is almost certain that the Gatlings could have turned the tables on the Indians.[15]

Machine Guns in the United States Army

In the following years, the Ordnance Department tested practically every gun that became available—the Maxim, the Lewis, and the Browning—and found them all wanting. As late as the Spanish-American War, several Gatlings were successfully used, but the Ordnance Department (literally and figuratively) stuck to its (old) guns. The case of the Lewis gun was particularly vexing. This was a lightweight gun invented by Samuel MacLean and O. M. Lissak, both Americans, but the gun was far from ready for production. They sold the patents to a company in Buffalo that asked Colonel Isaac Lewis, a recent retiree from the United States Army, to finish the design. By 1911 this was accomplished and the gun demonstrated to the United States Army, but nobody seemed interested. The Ordnance Department claimed that the gun was too flimsy to withstand prolonged service. The gun was tested (in 1912) as an air-to-ground weapon, but the War Department decreed that if a machine gun were to be mounted in an airplane, it would be the standard army machine gun, the Benet-Mercier. The fact that the Benet-Mercier had forty-five-centimeter-long feed and ejection chutes that dangerously interfered with the control of the aircraft did not interest the War Department, which ruled that it would be the Benet-Mercier or nothing.[16] Lewis took the gun to Belgium, where production was initiated and the gun purchased by the Belgian army. Then the British became interested and the gun was manufactured in England too. In the meantime the war started, and the demand for the Lewis was so large that the Savage Arms Company in the United States was subcontracted to produce them. The real or imaginary faults that the Ordnance Department found in the gun did not dissuade just about everybody else from buying it. This included the French, the Italians, the Russians, and of course the British, who adopted it in such a wholehearted manner and in such quantities that it saw service well into the Second World War, and it was generally accepted as a British design. What's more, during the First World War, because of its low weight, it was used extensively as an aerial machine gun, equipping both fighters and observation aircraft. But the weapon was not good enough for the United States Army. The U.S. Navy and Coast Guard adopted it, but upon arriving in France and being placed under the command of the army, the Marines were ordered to discard it and were reequipped with an inferior French machine gun.

Finally, under the pressure of the fighting in Europe, American factories started producing Browning and Vickers machine guns, but most of them never made it into the hands of American soldiers. It remained for General John Joseph Pershing, the commander of the United States forces in Europe, to supply the last ironic twist to this extraordinary situation. When the first Browning automatic rifles arrived, Pershing ordered them to be stored and not distributed to the soldiers. The reason was that they were so advanced and so efficient that Pershing feared if the Germans captured any of these rifles, they would copy the design and turn it on his troops. Admittedly, the design was quite advanced. The weapon remained in service until after the Korean War as the ubiquitous BAR (Browning automatic rifle), and for many years later as the FN (Fabrique Nationale) light machine gun.

This attitude toward firepower, and its place on the battlefield, changed rather abruptly between the wars and manifested itself at first during World War II. It is interesting to speculate (and might even deserve some research) if the present American propensity to use firepower whenever possible, and sometimes even when it is in fact counterproductive, has its origins in Ripley's twisted convictions.

In many respects the machine gun was a watershed in military technology, not so much because of its direct effects on combat (although these were far from negligible) but by later exposing the lack of willingness of both military and political leaders on both sides of the trenches to come to grips with the facts of life in a changing world. In a completely contrasting manner, it is possible that (at least for the time being) the military and political leaders did learn some lessons from history. With one controversial exception in August 1945, they managed to prevent the use of nuclear weapons in the struggle between the two superpower blocs, even in the presence of quite a few potential button pushers.

THE STORY OF THE NITRATES

Introduction

Nitrates, also called niter, are actually a group of three chemicals in which the element nitrogen is combined with oxygen and with potassium, sodium, or calcium. The first, potassium nitrate (also known as saltpeter), is an ingredient in the production of black gunpowder, the predecessor of modern smokeless powders. Until the perfection of modern explosives in the nineteenth century, black powder was also used as an explosive. Even today it is used extensively in various igniters for modern pyrotechnics, including military rocket motors. Potassium nitrate is a naturally occurring mineral in several regions, and when its consumption grew with the advent of firearms, it was also extracted from manure heaps and other organic refuse by mixing the organic matter with lime and letting it stand in the open air. Some was also scraped off walls and ceilings of cellars where it precipitated.

Being a critical ingredient in any gun- or powder-producing industry, during many periods the supply and collection of potassium nitrate were regulated by the state. In the beginning of the French Revolution, Lazar Carnot, a military engineer and mathematician as well as a member of the Committee of Public Safety, organized in Paris the production of arms for the mass armies of the revolution. Instructions for gathering potassium nitrate were printed and distributed all over France, and a special cadre of specialists in the production of powder was trained in Paris.

The two other nitrates, sodium nitrate ($NaNO_3$) and calcium nitrate ($CaNO_3$), are both used extensively in the chemical industry, and all three types can be converted from one type to the other by suitable chemical processes. The growing need for nitrates was eased when huge deposits of sodium nitrate were discovered in Chile in the middle of the nineteenth century. Sodium nitrate is actually named Chile saltpeter. It is essential to the production of nitric acid, an important chemical by itself, and is also a critical ingredient in the production of smokeless gunpowder and other explosives.

Nitrates, Fertilizers, and Ammonia

The problem of the availability of cheap nitrates was highlighted in the public awareness by another factor. Nitrates form the basis for another important group of products: artificial fertilizers.[17] At the beginning of the nineteenth century, the world was under the effect caused by the "Malthus Theory." Thomas Robert Malthus (1766–1834), an English economist and demographer, published a theory saying that since the world's population was growing in a geometric progression, while food production was growing at an arithmetic progression, the time was nearing when the world's population would simply run out of food. Consequently, there developed a scramble to increase food supplies, and one of the simpler ways considered was to increase the supply of fertilizer and make it cheaper. When vast amounts of sodium nitrate were discovered in its land, Chile quickly realized its position as the main supplier of nitrates to the world. The Chileans could write their own ticket regarding the price of nitrate exports. Reluctantly, the world continued to depend on the Chilean natural deposits. This dependence on imports for a critical and expensive imported chemical drove many researchers to try to synthesize nitrates, or for that matter any nitrogen-containing formula, by tapping the atmosphere's nitrogen. We live under a blanket of air that is composed of 80 percent nitrogen and 20 percent oxygen, but the nitrogen is a fairly stable gas that does not combine easily with other substances. It turned out that there was no easy or cheap way to combine the two into NO_3, which is the common ingredient in all nitrates.[18]

Fritz Haber was a brilliant German chemist who worked for I. G. Farben, the big German chemical cartel.[19] This immensely wealthy cartel hired promising chemists and allowed them to work on their own pet projects. Among these many projects were the development of aspirin, sulfa (an artificial antibiotic that was the predecessor of penicillin), atabrine (a malaria prophylactic), heroin and methadone, salvarsan (an early syphilis drug), and novocaine (a painkiller). Most of the developers of these products also won Nobel Prizes. Haber had already developed a simple test for pH (acidity) and completed original work on electrochemistry and thermodynamics. He obtained a grant from BASF, one of the companies in the I. G. Farben cartel, in order to find a better way to combine nitrogen with some other element, in other words an artificial nitrate. Within a year, in 1908, Haber perfected a process to produce cheap ammonia, NH_3, which was even better than NO_3 since it had more uses.

Literally overnight Haber became a national hero and a worldwide celebrity. In 1918 he won the Nobel Prize in chemistry for this discovery.[20] His fame did not help him, however, when the Nazis came to power in Germany. Being a Jew, he was fired from all his positions and was forced to emigrate to Switzerland, where he died in 1935. In three years, by 1912, Haber was named the director of the prestigious Kaiser Wilhelm Institute in Berlin. In the meantime, BASF strove to convert Haber's process into commercial enterprise, for the production of fertilizers, and in four years (also in 1912) achieved this under the direction of an engineer named Carl Bosch. In recognition of this feat, the process was later known as the Haber-Bosch process.[21]

The German General Staff and the Coming War

Meanwhile, war was nearing and the German general staff, although on the whole a well-educated group of people, suffered from a malady typical of many professional soldiers throughout history. They never bothered to inquire as to the source of their weapons, which formed much of the basis of their profession. It turned out that they had absolutely no idea that explosives production depended on nitrates, and while Germany had a process for producing ammonia, there was some way to go before this process could be geared to the production of explosives. In the meantime, Germany, like the rest of the world, was totally dependent on Chilean nitrates. The Chileans, being neutral in the war, were only too happy to supply Germany with all the nitrates she required. It may be added that there was also a large German colony in Chile, but that did not stop the Chileans from selling their wares to Britain, France, and the United States too. There was only one hitch. Chile is on the western side of the South American continent, facing the Pacific Ocean. The sea route to Chile is either through the Atlantic, near the Falkland Islands, which was a British navy outpost and a coaling station, and around Cape Horn, or through the Atlantic, around the Cape of Good Hope, and through the Indian and Pacific Oceans. From Germany's point of view, both routes were bad because most of these sea routes were effectively controlled by the British navy.

The German general staff made another mistake. Historically the Germans were most worried about the possibility of war on two fronts—simultaneously against the French and the Russians. This was in fact one of the holy writs of German strategic thinking—never fight both together. Count Alfred von Schlieffen, chief of the German general staff from 1891 to 1905, devised a plan in which German forces would cut through Belgium and Holland and race to Paris before the Russian army could mobilize. Then, after France was defeated, by moving the army on essentially internal lines with the help of the excellent railroad system, Russia in turn would be defeated. The plan depended on extremely rapid movement and conquest, and because most of the army was concentrated in the northwest, it was willing to accept some German retreats in the southwest. Unfortunately, because of possible logistics difficulties and so as not to surrender any German soil to probable French offensives, Moltke (Helmuth Johann, the Younger), the current chief of staff, removed forces from the right wing to bolster the left. He is often blamed for ruining the plan and robbing Germany of any possibility of a rapid victory. In this case, the general staff did not even consider what would happen if things did not go according to the changed plan, resulting in a longer war (remember Clausewitz and the inevitable "friction" in war), and ordered a total mobilization, including workers in key industries. This was done disregarding the possibility that these industries would have to continue to produce during the war.

A noted German industrialist named Walther Rathenau, the head of A.E.G. (Allgemeine-Elektrizitats-Gesellschaft, or Electric Corporation of Germany), tried to warn the general staff of the impending industrial disaster (they even mobilized workers in munitions factories), but the arrogant Prussian Junkers simply did not comprehend the problem. For one, they were irrevocably married to the concept of a short war.[22] Second, Rathenau was a civilian, and a Jew as well, so what did he understand about these matters? When the war started, Rathenau went to see the war minister (General Erich von Falkenhayn) and convinced him of the danger Germany was facing in this industrial war. Falkenhayn, who grasped the situation, started putting things in order and created a War Raw Materials Office headed by Rathenau. The most immediate problem was that of the explosives and nitrates. The most amazing fact discovered was that Germany went to war with a six-month supply of nitrates for her explosives and powder factories and with no way to replenish these stocks because of that accident of geography: Chile facing the wrong ocean. Rathenau called on Haber, who in turn called on Bosch, and they applied themselves to the problem.

In the meantime, Rathenau's predictions were proving right. The failure in the Battle of the Marne (September 6–8, 1914) dispersed any German illusions of a quick victory, and to those in the know it seemed that in a short while the German military would grind to a halt for lack of a simple chemical. The Germans did the reasonable thing from their point of view and attempted to force open the sea routes to Chile.[23] They moved a strong naval force (under the command of Admiral Graf Spee) from the Indian Ocean to the Pacific Ocean to impede British operations but ostensibly to act as commerce raiders. Because of poor judgment on the part of the British Admiralty, they sent a weak force to intercept Spee on the assumption that as an alleged commerce raider he would rather avoid battle. However, Spee was there to seek a fight. On November 1, 1914, the two forces met off Coronel (Chile) and the British squadron was annihilated. This German success did not last long. The British sent a powerful force to the area with orders to avenge the defeat. On the eighth of December 1914, Spee went around the Horn and attempted to take or destroy the British naval station in the Falkland Islands.[24] He ran into a British ambush, the force of Dreadnoughts (early modern battleships) looking for him, and except for one ship that was sunk somewhat later, all his force was destroyed there and then, without any British losses. Winston Churchill, who at the time was the first lord of the Admiralty, meaning in effect the supreme commander of the Admiralty and the navy, commented on this episode in his book about the First World War: "We do not know what were the reasons which led him to raid the Falkland Islands, nor what his further plans would have been in the event of success."[25] Churchill's remarks were quite logical. If the German squadron was to act as a commerce raider, it should have stayed away from sight of land; furthermore, as commerce raiders there was no point in bunching all these powerful ships together. It seems that the real reason for Spee's presence in the area, control of the sea routes in order to enable nitrate carriers to sail freely to Germany, never occurred at the time to anybody in England, including Churchill.

In the meantime, Bosch and Haber continued to work on the nitrate problem and finally in May of 1915, in the nick of time, managed to achieve nitrate production in sufficient quantities. Their six-month deadline of nitrates supply was extended by the unexpected capture of a large quantity of nitrates in the port of Antwerp when it was occupied in October 1914. Germany was saved, BASF found a gold mine, but the delay had most unpleasant consequences. It can be safely said that in the meantime it led to the introduction of gas warfare (which will be treated in the next chapter).

RADARS AND THE BATTLE OF BRITAIN

Introduction

Radar systems came into prominence during the Second World War and were the culmination of a long development effort carried out by many nations. The first device intended to locate ships by means of reflected radio beams was invented in Germany by Christian Hulsmeyer in 1904. He named his invention the telemobiloscope and exhibited it widely, demonstrating that it could locate ships up to five kilometers away. Seeking financial help to continue his work, Hulsmeyer approached the head of the German navy, Admiral Alfred von Tirpitz. True to form, von Tirpitz was not interested. Although it had undeniable military and commercial applications, the invention never caught on and for all practical purposes was forgotten. Radio technology slowly matured and in the thirties was advanced enough for scientists in many countries to think of practical applications for it. These included the United States, Japan, Germany, France, the Soviet Union, Italy, Holland, and England.[26] The Americans were actually the first to develop a working system and, after a first attempt in 1922, started again in 1930 and by 1937 had an experimental system for target detection and fire control mounted on a ship.[27] The French concentrated on radars for civilian uses, but their progress was slow and after the fall of France their plants contributed only to the German war effort. The Japanese scientists were quite advanced in this field, but poor management, lack of an effective guiding hand, and the rivalry between the services hampered the contribution of civilian scientists to the Japanese war effort and frittered away any lead they had at the start of the war.[28]

The Germans also made large strides in this new field, working on a variety of technologies in several research centers, and in 1938 started producing various operational systems for land and naval use. By that time the Germans were the most advanced technically and operationally. When the war started, they had radars mounted on ships for better (than optical) range measurements for gun-laying, land-based radars to locate approaching aircraft, and air defense radars for target acquisition at night. The air defense radars were supposed to locate the approaching aircraft and hand this information to the searchlight units that in turn were to illuminate the target and enable the antiair gunners to see it. While today it seems a most cumbersome procedure, we should remember that the technology at the time was not capable of pinpointing the target's location to actually aim guns at it, and the Germans improvised around this problem quite effectively.

For a long time the British were worried by the prospect of massive fleets of bombers coming over England, sowing destruction and death. They had already encountered this threat during World War I, when Germany launched indiscriminate bombings of English cities by means of bomber aircraft and zeppelins. Militarily, these bombings were useless, but under public pressure the government had to recall several squadrons of fighters from the western front to protect British skies. After the war, the Royal Air Force (RAF), although short of funding for bomber development, nevertheless studied the problem. This effort, coupled with the steady improvements in aircraft technology all over the world, made it clear to the British that their island could again be directly attacked, and this time with possibly devastating effects. The British took such a threat seriously and in 1928 started organizing an observer corps to warn of approaching enemy planes. However, the British understood that just seeing the aircraft was useless in itself, and so they also started thinking about a communications network for reporting such sightings.

After the rise of the Nazis, the British became increasingly worried. In December 1934 the British Air Ministry created a committee of outside experts to study the problems of air defense of Britain. This body was named the Committee for the Scientific Survey of Air Defence (CSSAD) and was headed by Henry Tizard.

Tizard was a nationally known figure. During his career, he served as the rector of the Imperial College of Science and Technology and as chairman of the Aeronautical Research Committee (similar to the American NACA). The CSSAD consisted of Tizard; two civil servants, H. E. Wimperis and A. P. Rowe; Patrick Blackett, who later was deeply involved in operational research and in 1948 won the Nobel Prize in physics; and Archibald Hill, a 1922 Nobel laureate in physiology. Tizard, Blackett, and Hill had military experience from World War I. Shortly afterward another member was added to the committee, Frederick Lindemann, also a scientist with previous military service and a personal friend of Churchill. Tizard's committee investigated various aspects of the problem of air defense and among other ideas asked Robert Watson-Watt, from the Radio Research Laboratory, to check the rumors about "death rays." This was no doubt influenced by H. G. Wells's book *War of the Worlds*, which was published in 1898 and included the idea of Martian invaders using some kind of a "heat beam" to vaporize and otherwise disable their opponents. Everybody agreed that if such a device could actually be made to work, it would be an awesome weapon. Watson-Watt asked his assistant to calculate the amount of radio energy required to raise the temperature of eight pints of water, at a distance of five kilometers and a height of one kilometer, by some four degrees centigrade. The assistant, being quite astute, immediately understood that the idea was to disable a man in an aircraft and after some quick calculation showed that the current technology was not up to such a feat. He remembered, though, an incident long ago, when passing airplanes had disturbed post office test broadcasts. Watson-Watt suggested to the committee that radio waves be used to discover airborne targets at a distance. The committee liked the idea, particularly since it promised much better results than the best concept it had devised up to then: the detection of aircraft by their sound. Some tests were performed, and after their successful conclusion (in February 1935), some initial funds were allocated by Air Marshal Hugh Dowding (then the commander of Fighter Command). The British government soon accepted the concept and toward the end of 1935 started constructing the so-called Chain

Home array of RDF (as radar was then called) stations.[29] As is apparent, the British came rather late on the scene in this field. The generally held misconception that the British invented radar probably stems from the important role of the Chain Home system of radar stations during the Battle of Britain, and its unqualified success in this role.

The British, however, realized that placing several radar stations is not enough of a solution and from the start were thinking in terms of an integrated system. They eventually constructed the stations in such a way that all the southern and southeastern shores were covered, with actual overlapping in some sectors. The antennas of these radar stations were nothing like today's relatively small dishes, enclosed in domes, but instead were huge—more than 100 meters (360 feet to be exact) high—affairs of open grid construction, and quite visible from a distance. Furthermore, the British understood early in the game that apart from the technical facet of the detection of aircraft, there was another, immensely more important organizational aspect necessary for the success of the idea. Watson-Watts's original suggestion also included the establishment of control centers to sift through the information coming from different radar stations and observers, collate it, and pass this information to the fighter squadrons. This in turn necessitated an extensive communications network. The British constructed this elaborate system, and this was probably the first modern C^3 (command, control, and communications) system in the world.

The Germans, who at this stage were still technically more advanced in the radar field than the British, did not develop such an integrated system. It was really more a problem of conceptualization than of technology. The Germans were thinking in terms of offense, not of defense. Consequently, they developed sophisticated electronic long-range navigation aids. These navigation aids were intended to help bombers find their way at night to darkened targets. It is interesting, nevertheless, to note that even in their offensive thinking the Germans did not go far enough. They never developed a long-range strategic bomber, nor did they develop a real escort fighter for long-range bombing missions.

The huge RDF antennas that were sprouting all along the shore could hardly be hidden, and soon the Germans became curious. In May 1939 they sent a zeppelin cruising along the English beaches. The zeppelin was loaded with radio receivers, and one of its passengers was Major General Wolfgang Martini, chief of the Luftwaffe's signal section, but after long hours of cruising and listening they heard nothing.[30] Most of the stations were shut down, but in any case the zeppelin was listening on the wrong frequencies. The more advanced German radar (the so-called Würzburg) operated at wavelengths of about fifty centimeters, and the Germans assumed that if the British had radars they would be operating on similar frequencies. The British, however, were operating with wavelengths of more than a meter, so the Germans heard nothing. They tried again in August with similar results, and so they concluded that whatever these towers were, they were not radars.

The Battle of Heligoland Bight

When the war started, it was the Germans who conducted the first successful radar-aided interception of bombers. On the third day of the war (September 5, 1939), the British sent a gaggle of bombers to bomb the wharves of Wilhelmshaven in the so-called Heligoland Bight. The Germans detected them soon enough, but because of poor organization it took too long for the fighters to be scrambled. By the time this was accomplished, the bombers were on the way home. Here the British made a cardinal mistake. For a long time the British believed that the Germans had no radar. So, blissfully ignorant, they continued conducting antishipping operations in the area, although in bad weather most of the time. December 18, 1939, was a clear and sunny day, and the British were attempting to finish off a German cruiser that had been damaged by a submarine attack. Nobody expected the RAF to attempt a raid on such a clear day, but in the early afternoon, two German radar stations discovered the approaching bombers at a distance of 120 kilometers. Even so, the intercept was almost botched because of poor communications between the radar stations, through the intervening command structure, to the fighter squadrons.[31] The fighters managed to take off just in time, and in the ensuing battle twelve of the twenty-two attacking Wellington bombers (twenty-four actually set out, but two returned with engine problems) were shot down, and three more crash-landed back in England. The Germans lost a few fighters. This battle brought about two important changes in the British thinking. Self-sealing fuel tanks were accepted as a necessity, and the British started to realize that unescorted daylight raids by bombers were a questionable proposition.

But the strangest thing of all happened on the German side. This battle was widely documented and reported, and because of conflicting claims about "kills," it was the subject of lively correspondence between the squadrons and the Reich's Air Ministry in Berlin. So all the details of this action were well known to the German command. Presumably they noticed, or should have, that the first alarm was given by the radar stations when the attackers were yet a long distance away. Such information should have been assimilated into the operational doctrine of the Luftwaffe. Nothing of the sort happened. One is almost tempted to conclude that the fighter squadrons, the army (whose radar it was at that time), and the Luftwaffe command were barely on speaking terms.

Eagle Day—The Battle of Britain

August 12, 1940, was Germany's official starting date for the effort to subdue the RAF and prevent it from contesting the coming invasion of England. *Adler Tag* (Eagle Day) was the result of the understanding that as long as the Royal Navy and the RAF remained viable fighting forces, such an across-the-channel operation would not succeed. Furthermore, it was deemed that the elimination of the RAF's Fighter Command would be the more important target. However, by listening to British radio transmissions during the preceding months, it was discovered that British pilots were directed from control centers on the ground, and apparently the Germans finally realized that those tall towers were the radar antennas.

Consequently, four of the first missions on August 12 were directed against these RDF stations and were generally successful.

One station was completely destroyed and three others damaged. The Germans tried to destroy the antennas, but the structures proved quite resilient. The Germans ignored the wooden shacks dispersed near the antennas on the assumption that these must be administrative buildings and not worth the expenditure of bombs. It stood to reason that the real equipment and control centers were deep in the ground under several feet of concrete. They were wrong. Besides, the British managed to have the three damaged stations properly functioning again in less than two hours. For the destroyed one, in a place called Ventnor, the British used deception. They quickly put in a transmitter (without the receiver) that worked on the correct wavelength. As an operational radar it was of course useless, since it could only make noise and see nothing, but the Germans did not know that. Their conclusion was that damaged British radar stations could be put back in operation after less than two hours and that therefore the four missions against them were in essence wasted. The effort should then be directed against the real targets—the airfields and aircraft of Fighter Command. Three days later, two more RDF stations were destroyed, but because of the overlapping of the stations and the fact that the Germans did not figure out the free (from observation) corridors, if there were any, the defense was not affected. As a result, after the attacks on August 15, Goering declared, "It is doubtful whether there is any point in continuing the attacks on radar sites, in view of the fact that not one of those attacked has so far been put out of action."[32] Later, while the fighting was still going on, the Germans were still troubled by the obvious success of the British interceptions. After the war, an expatriated British prisoner related that when interrogated about the interceptions, his answer was, "We have excellent binoculars." Whether the Germans bought this story or not remains unknown, but it is a fact that through the whole Battle of Britain, which lasted until September 15, the Germans made only several halfhearted attacks against the radar stations.

The German Defeat and Some Lessons from It

Be that as it may, this was probably the single, most telling technical blunder on the part of the Germans throughout the war. The defeat—and a defeat it was—in the Battle of Britain was undoubtedly the only thing that kept the Germans out of England, but it was a close-run contest. Fighter Command was on its last legs due to battle weariness and accumulating losses in pilots. Also, the communications network, probably the second most important element of the system, was near collapse because of peripheral bomb damage. If in addition the early warning network by radar had been decimated or even seriously damaged, the fighters would have had to spend countless additional hours in the air (thus wasting fuel, engine hours, and general wear and tear on aircraft and pilots) or risk getting caught on the ground. Under such conditions it is quite conceivable (and many authorities say that it is certain) that Fighter Command would have collapsed before the Germans finally gave up. Even so it was touch and go. Luckily for the British, on August 24 a lone German bomber dropped some bombs on London, apparently by mistake. Churchill ordered several retaliation raids on Berlin that caused minimal damage, but against the advice of Luftwaffe commanders, Hitler ordered a major offensive against London instead of continuing it against the fighter airfields. This happened on September 7, 1940. That was enough to give Fighter Command a breather. The intense fighting in the air continued for another week, but it was not directed specifically against Fighter Command and its airfields. The climax came on September 15 when the Germans were soundly trounced (sixty German to twenty-six British aircraft lost). Apparently that was too much for the Germans, but more important, it proved to them that Fighter Command was not at all beaten. Immediately afterward the Germans canceled Operation Sea Lion (the invasion) and started the so-called night blitz against British cities.

The importance of learning of and destroying the enemy's radar capabilities evolved into the varied and technically demanding field of electronic warfare. The British went as far as staging a special raid to lift certain components from a German radar station on the French coast. The successful raid took place in February 1942, and apart from the technical analysis of the components, the British gathered information on the rate of production of these radars from an analysis of the serial numbers stamped on the various components. In view of this success, a second such raid was planned concurrently with the Dieppe raid in August 1942, but the failure of the Dieppe operation prevented the radar operation from being carried out. A similar raid was staged by the Israelis in December 1969 during the war of attrition with Egypt, following the Six-Day War of 1967. A commando unit crossed the gulf of the Suez, captured and disassembled a whole (Soviet-made) radar station, and flew all of it by helicopters back to the Israeli lines. Finally, a little more than half a century after the German failure, Operation Desert Storm, in January 1991, was started with a massive attack against Iraqi radar stations and other command and control centers and led by stealth aircraft. Something after all had been learned.

STRATEGIC BOMBING AND THE SELF-DEFENSE OF THE HEAVY BOMBER

Introduction

The general problems of the allied strategic bombing effort during World War II have already been discussed extensively in the literature. Consequently, this section will not address the question of whether the bombing raids against Germany constitute a success or a failure, or the question of the return on the investment in these bombing raids, both in daylight and at night. Arguments on the subject still rage. It can be agreed that there were serious deficiencies in this air campaign, expressed in excessive bomber losses and the fact that these losses did slow the bomber offensive, particularly in the latter half of 1943 and until the long-range fighters became available. One question, however, was never addressed properly: where did the belief that the bomber will always get through originate, and how was it sustained? Was it, technologically speaking, at all justified at the time it was conceived, and if so, why did it suddenly collapse?

World War I Bombing Attacks

To understand the question of the problems of the long-range bombing in the Second World War, one must look at its roots during the First World War. Bombarding cities and their inhabitants is, of course, not a new or modern concept. However, in the past it was carried out only against besieged cities or where the city in question was directly in the line of advance. During World War I the Germans carried this approach to warfare one step further. The use of aircraft began as a means to observe the actions of the enemy but soon evolved to include the capability to drop explosives. This was extended to bombing targets behind the enemy lines, and the Germans also wanted to bomb England. England, however, was beyond the range of the then-current aircraft, and the Germans decided to employ zeppelins, lighter-than-air ships, which achieved their lift by being filled with hydrogen, a gas much lighter than air. Although slow, zeppelins had the advantage of long-range and large weight-carrying capability. In January 1915 the Germans attacked the south of England with zeppelins, and after several more raids fourteen German zeppelins attacked London at the end of May. Although the damage was slight, the British public was scandalized, as it appeared that the English Channel was no longer a barrier to an enemy's intrusions. When more raids materialized, fighter aircraft were sent aloft, aided by ground gunnery, and in a short while the slow, big, and extremely explosive airships were in effect swept from the sky. All told, there were some fifty raids and nearly six hundred people were killed.

The Germans then decided to continue this offensive by means of the Gotha bomber, which at the time was the most advanced machine they had for that purpose. This was a twin-engine aircraft, manned by a pilot and two machine gunners, that could carry 500 kilograms (1,100 pounds) of bombs and fly at an altitude of up to 15,000 feet. In April 1917, operating from Belgium (which was partially conquered by the Germans) the Gothas began attacks on the south of England, culminating in a daylight raid on London on June 13, 1917. Some 150 people were killed, half of them women and children. Several fighter squadrons went up, but all the Gothas returned safely. Public indignation aroused by the zeppelin raids now turned to panic, and the government had to do something.

The Smuts Report and the Creation of the RAF

General (later, in 1941, Field Marshal) Jan Smuts was one of the successful leaders of the Boers in the second Boer War (1899–1902), and his exploits and efforts in initially trying to prevent the war, and later at the peace negotiations, earned him the respect of the British. In 1910, with the establishment of the Union of South Africa, he became its minister of defense and in the beginning of the war participated in the campaigns against the Germans in Africa. In the beginning of 1917, Smuts went to London as South Africa's representative in the Imperial War Cabinet and later was invited by Lloyd George, the prime minister, to join the British War Cabinet. After the war, he continued in a distinguished political career both in South Africa and internationally. He helped create the League of Nations, was prime minister of South Africa, was instrumental in securing South African support for the British in the Second World War, was one of Churchill's advisers, and helped draft the United Nations Charter.

In the wake of the Gotha raids, a knowledgeable and publicly accepted figure was sought to advise the government on this matter. In July 1917 Smuts was appointed to inquire into the matter of enemy raids on England and what could be done to thwart the threat, and within five weeks he produced the so-called Smuts Report. It was far thinking and included several important ideas, but most important, it planted the seeds for the bombing raids of World War II. Smuts concluded that an effective defense of the country against marauding bombers was in essence impossible, and thus the only counter to the threat would be the reciprocal bombing of the enemy's territory, conducted on a more massive scale. This in effect would change warfare, making air war the dominant form, with the land and naval battles of secondary importance. In order to achieve the desired effectiveness in this new kind of warfare, Smuts called for the creation of a new fighting service, which would absorb the Royal Flying Corps (which was part of the army) and the naval air arm, and would constitute an independent air force. Smuts reasoned that air war was becoming too specialized, and thus the new service had to be run by specialists, under its own air staff and under the control of an air ministry, which would have to be created. Although as noted Smuts had extensive military experience, it is interesting to speculate if he would have made the same recommendations if he had also had technical or flying experience.

Smuts's proposal was strong stuff, but surprisingly the politicians accepted it. The RAF was created on April 1, 1918, but by then bombing operations against Germany were already in progress. What nobody seemed to notice was that by that time the technology of fighter aircraft had advanced to such a degree that the Gothas went the way of the zeppelins and were prevented from daylight raids over England. They reverted to night bombing, but many of the bombers were lost in night landing accidents at their home bases. Nor were the results of these raids very spectacular. All told, the Gothas killed some 850 and wounded less than 2,000. Compared with what was happening on the western front, where such figures were sometimes the result of a single morning's skirmish, the civilian casualties and damage to property were miniscule. The Americans, after entering the war, attempted daylight bombing of German positions and soon learned that this could not be accomplished safely without escorting pursuit planes.[33] The RAF also started bombing military targets in Germany and until the end of the war dropped some 550 tons of bombs, but these raids too had only a minimal effect on the outcome of the war. The RAF ordered a four-engine bomber capable of carrying one and a half tons of bombs all the way to Berlin. Only three of these were delivered before the war ended, but later, in the thirties, this design served as the basis for the new generation of heavy bombers, culminating with the Lancaster, which could carry up to ten tons of bombs. However, after the end of the First World War nobody wanted to spend money on bombers, and the RAF justified its existence by providing an easy and very cheap way to police the expanses of the empire in the Middle East and Asia against native tribes.

The Theories of Victory through Bombing

In 1921 Guilio Douhet appeared on the scene. Douhet was an extremely outspoken Italian officer. Before the war, although an artillery man, he was given command of Italy's first air unit, which in 1911 carried out the world's first aerial bombing, in Libya, a fact that may have colored his later thinking. During the First World War, Douhet criticized the Italian conduct of military operations to such an extent that he was court-martialed and jailed (an even worse treatment than that of Billy Mitchell). Shortly afterward, however, the investigation of the costly Italian defeat at Caporetto proved him right and he was exonerated. After the war, Douhet was promoted to general, and in 1921 he published a book titled *The Command of the Air* in which he predicted that in a future war the defense might again prove superior and the war would once more degenerate into trench warfare. His solution to this was the establishment of an independent air arm, which would bomb the production and transportation centers of the enemy and demoralize the population to such an extent that they would force their government to sue for peace. This was to be achieved by well-armed bombers that would brush aside any opposition and fight their way to the target. It is not clear whether Douhet ever saw Smuts's report, or even knew about it. But in any case, he arrived at his conclusions from a different direction, so he may be granted the benefit of the doubt. Some authors have taken Douhet to task for not envisioning radar and other electronic devices, but more significant is the fact that Douhet did not correctly assess the development of the tank, airborne troops, and the modern fighter airplane.

On the other side of the ocean, Billy Mitchell (1879–1936), who by the end of the war was in charge of American air operations on the western front, was calling for a strong air arm. There is no question that Billy Mitchell's ideas were his own, because Douhet wrote in Italian and it took time for his book to be translated and his ideas to circulate. On the other hand, it is almost certain that the development of the strategic bombing concept, and the heavily armed B-17 in the United States, were influenced to some extent by Douhet. In any case, this was a period of great advances in aviation, pushed along by a civilian market that was interested in passenger planes. While bomber aircraft and passenger planes are very different in their overall design, the requirements for both are quite similar. They are to carry maximum payload at maximum speed to maximum range. This is a good start on the requirements of a bomber. Consequently, at a time when fighter aircraft were predominantly biplanes with fabric-covered wings and open cockpits, passenger planes and bombers sported mono-wings, metal fuselages, and retractable undercarriages. In exercises held in the United States in 1934, time after time the bombers easily penetrated to their targets. In those days, before the advent of radar, it was deemed that fighters should have some 40 to 50 percent speed advantage over the bombers in order to be able to catch up with them. But at that time the bombers were more or less of equal speed to the contemporary biplane, fabric-covered fighters. Apart from having the troops practice their trade, military exercises are intended to prove or disprove novel theories of warfare, and the conclusions from these mock attacks were rather unmistakable: the bomber could always get through. These exercises only confirmed an attitude that was already prevalent on the European continent. In November 1932, in a speech in the House of Commons, Stanley Baldwin, the former (twice) British prime minister[34] stated in part: "I think it is well also for the man in the street to realize there is no power on earth that can protect him from bombing, whatever people may tell him. The bomber will always get through, and it is very easy to understand if you realize area and space."[35] The speech was quoted in the *New York Times,* appearing as a heaven-sent gift to the big-bomber proponents in the States. Now came the 1934 exercises that conclusively proved the point.

Ironically enough, in England the speech had the opposite effect. As related earlier, two years after Baldwin's statement the Tizard Committee for the Scientific Survey of Air Defence was formed, and one of its more profound acts was to approve the establishment of the British radar chain and the whole elaborate system of command and control of Fighter Command, which eventually saved Britain. There was another ironic twist to this episode. When the dimensions of the German rearmament became known, there was a public outcry for more airplanes to match the Germans. The public did not understand the finer points of difference between bombers and fighters. They just wanted larger numbers. The British government also realized that the RAF would be hard-pressed to bomb every airfield and every other worthwhile target in Germany and that diverting some money to prevent the Germans from doing the same thing to England might be a worthwhile idea. Since fighter aircraft came at a considerably lower price tag than bombers, it was easy for the British government to keep the public happy and fulfill an urgent defense need.

The Germans tried unescorted bombing missions in Spain and quickly learned that this was costly. When World War II started, the British tried such missions several times and after some terrible losses found they were not immune from the realities of daylight bombing either. Then came the Battle of Britain. This was essentially strategic bombing, although not aimed at breaking the population's morale. The thinking was sound enough. By bombing Fighter Command's facilities, the Germans tried to force it to come up and fight. The bombing and the fighter combat would destroy Fighter Command, or weaken it sufficiently, so that it would not be able to oppose an invasion. Although the theory was sound, the execution was poor. The Luftwaffe did cause considerable damage and came close to accomplishing its objective, but these successes came at the price of prohibitive losses to itself and so they were forced to abandon the attempt.

The story of the German failure is somewhat involved and has a significant bearing on the later American thinking. The German military was thinking predominantly of a quick victory on land. That meant close air support for the ground forces. Regular horizontal bombing did not seem to be accurate enough and thus the idea of the dive-bomber was born, advocated strongly by Ernst Udet, at the time serving as inspector of fighters. Initially the idea was not popular, but over time General Walther Wever, the Luftwaffe's first chief of staff (1933–1936), warmed to it and eventually the Stuka (Ju-87) was born. But Wever was a bright officer. He realized that the real objective of German expansion would be the Soviet Union. This did not require political astuteness. Anybody who had read Hitler's *Mein Kampf* would have reached the same conclusion, but Wever also understood the military ramifications of such a campaign, at least from the air arm's point of view, namely the need for a long-range heavy bomber, which he openly called the "Ural Bomber." Such a bomber could also be very effective against Britain, both by bombing it directly and by operating against her seaborne supply routes. So from 1935 Wever pushed for the development of such a bomber, and contracts were actually let to Junkers and Dornier to develop aircraft such as the Ju-89 and the Do-19. This was done in the face of opposition from both Hitler and Goering, who believed more in close support of land operations. In the beginning of 1936, prototypes of both aircraft were flying, although it turned out that they were badly underpowered. There were not yet suitable engines for such bombers, but it was felt that this deficiency would be rectified in time. However, in June 1936, Wever was killed in a crash. The mediocre performance of the aircraft—but more important, the death of its champion—ended the project. After the Battle of Britain and when the Battle of the Atlantic was warming up, the Germans tried to speed the development of the He-177 (a four-engine airplane but turning only two propellers) and also converted the FW-200, a rather successful airliner, into the long-range reconnaissance-cum-bomber. Like many such jury-rigged solutions in highly technical fields, neither was successful.

American observers in England took all this in, both the British and the German failures, but persisted in their belief that the B-17s would get through, even in the face of fighter opposition. The reasoning behind this belief was based on several technical points. First, the B-17 was the first bomber designed to take care of itself. It was capable of operating at high altitudes—up to 25,000 feet in the early models. This was above most of the flak (antiair artillery), and at these heights fighter aircraft of the period were less maneuverable. Depending on the model, it carried between ten and fourteen machine guns with enough crewmen to man almost all of them simultaneously. That is why it was called the "Flying Fortress." The bomber groups were to be flown in such a formation (the staggered boxes) that at any time all guns on all bombers had a clear line of fire and all bombers could drop their bombs simultaneously. Finally, the B-17s were equipped with the most advanced bombsight (the Norden bombsight) in the world, and it was felt that the precision afforded by this bombsight would make short work of German industry.

Worldwide Advances in Fighter Technology

While without doubt the Allied thinking and approach to strategic bombing was considerably more advanced than the German preparations, both the British and particularly the Americans made one cardinal mistake. The mistake evolved from the subconscious misconception that technological progress is slow in the extreme and that future tactical considerations can be based on today's technology. This "Still Photo Syndrome," the assumption about the slowdown of technological progress in any given field, was quite prevalent in the past and is very common even today. Although it is common to nontechnical people, its most disastrous consequences occur when it is practiced by high-ranking and influential military people.

To some extent one can follow the effect of this thinking from the chronology of development of the various modern aircraft. The B-17 was designed in 1934, the first prototype flew in July 1935, and it was first delivered to the army air force for testing in 1936. Originally it was to be armed with .30″-caliber machine guns, but these (or at least most of them on each airplane) were quickly replaced by .50″-caliber machine guns. At that time most of the operational fighter aircraft in the world were fabric-covered biplanes armed with two rifle-caliber (.303″, .30″, and 7.92 millimeter) machine guns as shown in the following table:

Table 3-3: Types of Fighter Aircraft, 1933–1937

Airplane	*Country*	*First Flight*	*Armament*
He-51	Germany	1933	2 x 7.92mm
Ar-68	Germany	1934	2 x 7.92mm
Boeing P-26	USA	1933	2 x 0.30 inch

Bristol Bulldog	UK	1930	2 x 0.303mm
Hawker Fury	UK	1930	2 x 0.303mm
Gloster Gladiator	UK	1934	4 x 0.303mm
Fokker	Holland	1936	4 x 7.92mm
He-112	Germany	1935	2 x 7.92mm + 2 x 20mm
Polikarpov I-16	Soviet Union	1933	2 x 7.62mm + 2 x 20mm
Moran-Saulnier 406	France	1935	2 x 7.5mm + 1 x 20mm
Nakajima S-96-2	Japan	1936	2 x 7.7mm*
Nakajima S-97	Japan	1937	2 x 7.7mm

* The Japanese airplanes were metal mono-wings, but their armament still left something to be desired. The French airplane was also a modern metal mono-wing.

As can be deduced from the above table, with the exception of the Soviet Union, more heavily armed fighters started appearing only after the B-17 was already flying. The first Me-109 prototype flew in May 1935, the first Hurricane flew in November 1935 and the first Spitfire in March 1936.[36] Also, the first Zero flew in April 1939 and the first P-40 flew in October 1938. Furthermore, the first Me-109s were armed with two 7.92-millimeter machine guns, while the Spitfire and Hurricane carried eight .303 (7.69-millimeter) machine guns. However, there was already talk of going to bigger calibers. The general characteristics of these new aircraft, as well as details of their armament, were known to everybody. All manufacturers were desperately trying to peddle their wares to anybody and advertising their capabilities quite openly, including in various international competitions. They were also discussing improvements in high-altitude performance.

As noted above, the B-17 was finally armed with the .50-caliber machine gun, and although its basic design dated from World War I, it was an excellent weapon.[37] It also had a considerably longer reach than the rifle-caliber machine guns then in service on the fighters (even with the additional speed of the bullet imparted to it by the velocity of the aircraft), and it fired a bullet that was almost four times heavier. Thus, by every criterion the firepower of the B-17 was better than the firepower of the fighters extant when it was designed, and it probably could hold its own against the first fighters of the new generation.

The rifle-caliber machine guns, however, had many shortcomings in air-to-air combat, and when the B-17 made its debut, international standards for fighter armament were already changing and all major powers were thinking of something heavier. The British went to twenty-millimeter cannon (sometime retaining some rifle-caliber weapons), and the Germans armed their fighters first with fifteen-millimeter and later with twenty-millimeter cannon. When this started happening (in the time period of 1937 to 1939), the weapons carried by the B-17, and consequently its whole concept, were totally outclassed.[38] Thus, even the design of a special model of a B-17 that did not carry bombs but only guns did not help much. It is not known whether anybody ever considered arming the B-17 with twenty-millimeter weapons, but it in fact could not have been done because of the added weight of the guns and their installation (even with single guns instead of twins in each turret) and the weight of a reasonable amount of ammunition. This whole problem of relative firepower was well understood both to the airmen and to the aircraft designers. When the specifications for the new bomber, the B-29, were put forward in the early forties, it was to carry only twenty-millimeter cannon with quite advanced aiming systems. What would have happened if the war had dragged on and B-29s were sent to the European theatre is moot. Piston-powered fighter aircraft could not carry much heavier guns without losing too much of their performance, as actually was discerned in later models of the Me-109. One possible solution to this problem, which was already adopted against the B-17s, was air-to-air rockets, but further speculation will not be made here.

The Choice of Targets

While the lack of appreciation for the evolution of firepower was a first-class mistake, it appears that the designers of the air campaign made another, more subtle mistake. In very general terms, the mission of strategic bombing was to cripple the enemy's industry, preferably by destroying some nodal points. Such attempts were made to hit crucial industries that would have paralyzed other production systems, with varying degrees of success. The raids against the ball bearing plants in Schweinfurt and the raid against the Ruhr valley dams were such special missions.[39] But the people responsible for deciding on target priorities apparently did not grasp the degree of integration of modern industry. The *Strategic Bombing Survey* of 1945 stated that a better targeting strategy could be achieved by consistent bombing of the electrical power generation and distribution system.[40] This would have paralyzed the rest of the German industry without actually having to bomb it out. Such a strategy was later carried out against the synthetic oil industry (after D-Day) and did achieve the expected results.

THE IDF (ISRAELI DEFENSE FORCE) CONFRONTS THE SAGGER MISSILE

Probably the classic example of a technological failure leading to a battlefield surprise in modern times was the appearance of the Sagger missiles in the hands of the Egyptian and the Syrian armies during the 1973 Yom Kippur War. Admittedly, the technological failure was accompanied by a doctrinal failure. It is possible that if not for that concurrent doctrinal failure, the technological failure and the resultant surprise would not have been brought to light so sharply, or might have been completely prevented. On the other hand, it is almost certain that a thorough analysis of the impending technological facts could have prevented the problem that developed on the battlefield, or at least ameliorated it to a large degree, either by devising some timely countermeasures, or more likely by reinstituting the correct doctrine, suitably adapted to the new conditions. As we shall see, a doctrinal change alone, no matter how clever, cannot long compensate for a technological failure. But in this case, such a change would have given the IDF a breathing spell to come up with a technological answer and probably would have prevented the so-called October Earthquake.

In order to better understand both the doctrinal and the technological failures and the link between them, we must start with some history. During the wars between the Arab countries and Israel, the Arab military forces did not really distinguish themselves. They failed in 1948 against what was basically a half-trained militia. In 1956 thousands of Egyptian soldiers doffed their shoes to run better, and many of them died in the desert. The 1967 war was a repeat of 1956 but on a grander scale, with Israel fighting a simultaneous war against four Arab countries. Again, photographs of piles of abandoned shoes lying around in the Sinai desert appeared in the press.

In 1956, during the heyday of close cooperation between France and Israel, the IDF bought a quantity of the French-designed SS-10 antitank missiles. The SS-10 was the first of its type in the world and constitutes one of the most original (among very few in the modern era) French contributions to the science of armaments. It was an optically guided, wire-controlled missile with a range of two thousand meters. The operator fired the missile and by means of a small joystick guided it to its target. In order for the operator to see the small missile (with a body diameter of sixteen centimeters) at the longer range, it was equipped with a bright red chemical flare at the tail.

Apart from the necessary good visual/motor coordination required of the operator, there was another requirement. When fired directly at the target, the flare's image was large enough to completely mask it. Thus, a special technique was necessary to overcome this problem. The operator guided the missile off course and came at the target from the side. Doing so demanded real-time, accurate determination of the relative ranges of the missile and the target. The same difficulty, and essentially the same solution, was inherited by the SS-11 and later by other missiles designed on the same principle in many countries, culminating in the Soviet Sagger. All these missiles required constant training for the operators, under varying conditions, in order to maintain reasonable proficiency. Still, the expected hit rate was around 10 percent.[41] Israeli operators achieved somewhat better results, mostly because they were handpicked personnel and highly trained.

In 1956 the IDF used the SS-10 successfully, though to a limited extent, against the Egyptian army. It should be pointed out that there was no question about the effectiveness of the missile when a hit was registered. During the Six-Day War (1967), the IDF encountered the Snapper missile, which was an early Soviet attempt to develop an antitank missile. However, because of the rapid collapse of the Egyptian army in that war, the Snapper was almost never used, and in the flush of that victorious war nobody paid it much attention. The IDF continued to keep some antitank missiles in its inventory, mostly Soviet ones taken as war booty, but the effort was only halfhearted. Instead the IDF fostered the tank gun as the best antitank weapon and admittedly brought tank gunnery to a pinnacle of efficiency at which other armies could only marvel.[42]

Understandably, the IDF became rather self-confident. This in itself is a positive, even praiseworthy trait, but unfortunately the IDF went one step further. Adapting doctrine to changing situation is a sine qua non of healthy military establishments, but there are several basic tenets of military theory that can be tampered with only at one's peril. One of these beliefs is the basic need for cooperation between the various arms, in this case infantry and armor. This imperative for the cooperation between (mechanized) armor and infantry in the attack was discovered in World War I and reaffirmed in World War II. However, because of certain deficiencies stemming from its history, the IDF was occasionally rather cavalier in its treatment of established (and proven) theory and tended too often to apply previously obtained empirical results to current problems without considering the overall picture. In this case the latest empirical results were those of the 1967 war.

Consequently, this second defeat of the Egyptian army in mobile warfare brought another change in the thinking of the IDF. It is generally accepted that tanks must be accompanied by infantry. The infantry's role is to take care of enemy infantry who may hamper the armor's operation by the use of various antitank weapons. Since tanks by themselves cannot handle infantry too well, they need friendly infantry to do this for them. The armor in turn will take care of the enemy's armor. However, there is a problem. The tankers sit behind four to eight inches of steel, while the infantry is running around wearing only a helmet and a thin shirt.[43] Even though the tank is a very distinct target, and in a way draws a lot of fire, this discrepancy in personal protection results in a great number of infantry casualties.

The piles of forlorn shoes on the Sinai dunes bred the idea that maybe the IDF did not need infantry to accompany the tanks. The reasoning was that the minute the tanks showed up the enemy's infantry would run away again, so why bother with friendly infantry that would only get in the way and might yet get shot up? The stage was set for a monumental blunder.

The Sagger missile is a three thousand–meter range, wire-guided, first-generation (CLOS—Command to Line of Sight) antitank missile developed in the sixties by the Soviet Union. Its Russian designation is AT-3 "MALYUTKA," and although it was the successor of the Snapper, like all first-generation antitank missiles, it was actually a descendant of the SS-10. This missile was first seen in the 1967 May Day parade in Moscow, and its picture appeared in many professional magazines throughout the world. It was supplied to all the Warsaw Pact countries and other political allies of the Soviet Union, including many Arab countries, replacing the Snapper in their arsenals.

The first information about the Saggers appeared in the open literature toward the end of the sixties. A description of the missile was published in *Ma'arachot*, the IDF's official magazine, in July 1970. In September of 1971, the same magazine published an article about antitank warfare doctrine in the Egyptian army and speculated that the Egyptians were probably being equipped with the Sagger as a replacement for older systems. In fact, immediately after the Six-Day War, the Egyptian army started looking for a way to overcome the Israeli superiority in mobile warfare, which they knew they could not match directly. The conclusion of these deliberations, arrived at with the aid of their Soviet mentors, was that some form of novel technology might outdo the qualitative superiority of the Israeli manpower. In the beginning of 1973, IDF intelligence knew of these Egyptian efforts but considered these reports highly classified, and so the information was not made common knowledge to all regular and reserve formations.

On October 6, 1973, the Egyptian army crossed the Suez Canal and the war started.[44] A hasty armored counterattack launched by the IDF, but without infantry support, was repulsed with heavy losses to the tanks. To everybody's amazement (no doubt including the Egyptians themselves) the Egyptian infantry did not run away the moment the Israeli tanks showed up, as they had in past wars, but instead stood their ground and fought with their Saggers and RPG-7s (a bazooka-like weapon). By the time the IDF realized that the operational doctrine for the employment of this missile substituted quantity for quality and was further supported by nearly unlimited numbers of cheap RPG-7s, it was almost too late.[45]

In actual count the missiles caused less damage than the Egyptian tank guns, the traditional antitank weapon. But the impact of the "surprise" of the "new" missiles, coupled with the overall strategic surprise of the war, magnified the missile's contribution to the IDF's initial difficulties. What's more, it enabled the infantry to stand their ground.

It is difficult not to make comparisons between the IDF's debacle on the banks of the Suez and the French defeat in the Battle of Crecy. Although separated by a continent and more than six hundred years of tactics and technology, the similarities are obvious. In both cases an armored force attacked without the proper preparation—a well-trained infantry force equipped with just the right weapon to repulse this kind of attack. In both cases the attacking forces had already met (and in the Israeli case even used) the said weapon, but its efficacy somehow did not register in the minds of the leaders. Finally, in both cases this lack of intelligence (in both accepted meanings), overconfidence, and arrogance brought about the same results.

The final irony was that at this junction, the Israeli failure caused the whole world to sit up and take notice of the capabilities of the antitank missile. It might have even influenced the thinking in the United States and the development of long-range precision-guided munitions for use in a ground war. It is possible that if by a combination of forethought and circumstances, the IDF had prevailed from the start against the missiles and antitank rockets, the worldwide development and introduction of these and other "smart weapons," and their technologies, would have been slower.

THE EFFECTIVENESS OF THE PATRIOT ANTIBALLISTIC MISSILE SYSTEM

Introduction

On September 17, 1980, Iraq attacked Iran. Considering the sorry state of the Iranian army after the Khomeini revolution and the subsequent purges, the Iraqis expected a quick and easy victory. However, the Iraqis did not achieve this quick victory, and the Iranians had time to mobilize their vast manpower resources. The discrepancy in population sizes (about 57 million in Iran compared with about 20 million in Iraq) and the complete disregard for casualties caused the war to drag on. While the ground warfare swayed back and forth near the border, in March 1986 Iran started firing Scud missiles at Baghdad. The Scud-B has a range of about three hundred kilometers and Baghdad is somewhat less than that from the Iranian border. The Iraqis too had Scuds, but it so happens that Tehran is more than five hundred kilometers from the border. Thus the Iraqis had a much greater challenge. They sent some aircraft (which in fact could carry a much larger payload), but the psychological impact of the missiles on the Iraqis was tremendous.

The War of the Cities

Under the circumstances, the Iraqis were at a considerable disadvantage and simply had to find a way to increase the range of their missiles. They chose to do so by combining two fixes. First, they reduced the size of the warhead from about six hundred kilograms to about three hundred kilograms, enabling a given amount of fuel to impart higher velocity to the whole missile. Second, they increased the volume of the liquid fuel tanks of the missile, enabling the motor to fire longer and thus attaining even higher velocity of the missile at burnout. Higher velocity enables longer range the same way that throwing harder gives an object higher velocity and longer range. Another issue facing the designer is whether, at some stage of the trajectory, to separate the warhead from the rest of the missile. If separated it is possible to achieve better accuracy, but this requires a separation system, with its added weight and complexity. On the other hand, since in many cases missiles tumble freely when out of the atmosphere, a nonseparated system will be more strongly acted upon by aerodynamic forces when reentering, resulting in diminished accuracy. Furthermore, missiles have complicated aerodynamics. Without going into too technical an explanation, suffice it to say that for an object like a missile to remain stable when traveling in the atmosphere—meaning that the nose points more or less forward—certain conditions must be fulfilled. One requirement is that the center of gravity be at a given point along the length of the missile. Reducing the weight of the warhead moves the center of gravity rearward. This is permissible but within very small limits. Changing the length of the missile (by changing the size of the fuel tanks) throws these very exact calculations off and requires an extensive recalculation and testing of all the aerodynamic parameters of the missile.

In the case of the Iran–Iraq war, because of the increased aerodynamic loads caused by the higher velocity, the stresses on the longer-range Iraqi missiles (named El-Hussein) were too great and they tended to break up when reentering the atmosphere. This of course played havoc with the accuracy of the missiles. But since the Iraqis intended the missiles to be terror weapons, fired against a sprawling metropolitan area and not against a military target, this was not a real problem. In a way this was similar to the German way of thinking in their employment of the V-weapons.

The breakups of the missiles were widely reported, including coverage on Iranian television,[46] so in essence the information was available in the west and should have been assimilated by those concerned. But at the time the missile breakups were considered to be no more than a curious phenomenon, attributed to some design flaw (which it really was) on the part of the Iraqi engineers and their Russian or German mentors, and no more notice was taken of the problem. Furthermore, American military leaders did not consider the Scud an effective military threat, and it was almost ignored. Of course, political considerations had not yet entered the picture.

The 1991 Ballistic Missile Attacks

One of the oldest ploys against ground-to-air fire is the use of decoys and various other countermeasures. The earliest such device was "window," used against German radar in World War II and described in chapter 5. A disintegrating missile also constitutes such a countermeasure by presenting a number of targets to the defending radar or missile. While there are ways and means to overcome such countermeasures, the exact parameters of the problem must be studied and the proper ways to respond to these countermeasures discovered and optimized. Even with unlimited resources this takes time.

On January 18, 1991, the Iraqis started firing El-Hussein missiles toward Israel and Saudi Arabia. Patriot missile batteries were rushed in, and interception attempts were made in both Saudi Arabia and Israel. At the time, it was reported that the Patriot missiles did succeed in intercepting some of the Iraqi missiles, but later it was shown that most if not all of the intercepts were actually carried out against broken-off parts of the reentering missiles. Only after the war was the information about the breakup of the Iraqi missiles over Tehran correlated with the breakup of the missiles over Saudi Arabia and Israel and to the failure of the interceptions.

IRAQ IN THE PERSIAN GULF WAR OF 1991

So much was written about the various failures of the Iraqi leadership during the war of 1991 that it is doubtful that anything new can really be added. However, the question remains: did the Iraqis understand the weapons they were going to face or were they surprised?

To start with, the Iraqi leadership was generally aware of modern weapons technology. They were acquiring and modifying ballistic missiles; they were in the process of enlarging the weapon industries; the Iraqi electronics industry and its efforts in this field were the most advanced of all Arab countries; they had nuclear ambitions; and they spent a lot of money on all these ventures. So the leadership had some understanding, or at least a belief, that investing in weapons was worth the expense. Although in many Third World countries weapon acquisition is often a matter of prestige, intended to enhance the ruler's position, this was not so in Iraq. Iraq had just emerged from a long and costly war, which historically speaking did not really end, so we can assume that some of these acquisitions were intended to be used eventually.

Unless they constitute some major technological breakthrough, specifications of modern weaponry are rarely kept in locked safes. The weapons market is a buyer's market, and there is always cutthroat competition to sell the wares. There are dozens of glossy magazines in this field, and the number of weapons exhibitions and conferences throughout the world is more than any one person could attend. In short, if you want to keep up to date on the latest in this field, all you have to do is subscribe to several good magazines and attend several exhibitions each year. No doubt the Iraqis did this. Furthermore, being repeat customers with a lot of money, the red carpet was rolled out for them any time they came to do business, and they were probably shown equipment reserved only for good customers. Even if they did not buy everything shown to them, they would have had a fairly good appreciation of what was state of the art in modern weaponry, particularly in such countries as the United States, the U.K., and France, let alone the Russians who were longtime suppliers to the Iraqis. However, this kind of information was usually known at the middle levels, and only selected items were percolated to the higher levels of authority.

In addition, most of the American weapons concentrated in the Gulf area during Desert Storm were fairly old. The Hellfire missile was originally introduced in 1970, and the Tomahawk cruise missile in 1981. Even the F-117 stealth fighter had already participated in the fighting in Panama in December 1989 and was extensively described in the press. Admittedly, it is doubtful if at this late stage the Iraqis could do much about these weapons, but on the other hand the quantity and quality of this arsenal should have told them something. The Japanese made a similar mistake in 1941 in their appreciation of the American industrial potential (although Admiral Isoroku Yamamoto himself had his doubts about the advisability of the war).

One potential explanation for Iraqi recalcitrance is that they did not believe the coalition would actually go to war. Another possibility is that they believed some published reports that all these weapons were too complicated and thus unreliable. But the simple, most likely explanation is that even after buying all these fancy weapons they did not really understand modern technological war.

The Iraqis considered the desert their own home turf and believed that the coalition forces would have difficulties in such terrain. They considered off-road travel in the desert as verging on suicidal because of navigation difficulties. They were tied to the roads and assumed that any attacking forces would face the same limitations.

Desert navigation (based on similar principles to sea navigation) had already been developed before World War II and later used by the long-range desert groups in the Western Desert.[47] The Iraqis' apparent lack of awareness of these facts is almost excusable. However, they were either not aware of or did not understand GPS (Global Positioning System—the satellite navigation system) and the total freedom it confers. Handheld GPSs were by then available for commercial and sport use, but it seems the Iraqis were not aware of that either. Admittedly, the Americans were not equipped with these either, but once in Saudi Arabia they realized the problem and bought all the available stocks in the United States.

The GPS is just one example of a basic technology that is now instrumental in winning a war. There were of course many more, and their sum had a synergistic effect. What happened in Desert Storm was not just a surprise. This was so different from the Iraqi experience in the war with Iran that they were completely overwhelmed, both physically and psychologically, and this led to the quick collapse of most of the Iraqi army in 1991.

4 Misunderstanding Available Technology

The tank was a freak. The circumstances which called it into existence were exceptional and are not likely to recur. If they do, they can be dealt with by other means.

Major General Sir Louis Jackson, 1919

THE FAILURE OF GAS WARFARE IN WORLD WAR I

Introduction

The use of organic and synthetic man-made substances to cause sickness or death to the enemy is as old as warfare itself. Catapulting the carcasses of dead animals into a besieged city and poisoning wells were practiced as often as setting fire to a building or a field to force out the occupants. One of the more famous such fires was the one the Saracens (Moslems) under Salah A Din (Saladin) ignited in the Battle of Hittin near Tiberias in Palestine, against the crusaders, on July 4, 1187. This battle was fought at the height of summer, in one of the hotter parts of the country, and apparently on a day with an east wind blowing from the desert. With the fire and smoke in their face, and lacking water, the crusader army was totally annihilated. The advent of the chemical sciences brought new possibilities, but by then the nations of Europe had ruled out the use of noxious substances as means of war (in the Hague Conferences of 1899 and 1907). Winston Churchill relates that there were suggestions to use noxious gases and smoke in World War I, but the British refrained from using this course of warfare first.[1]

The lack of nitrate, which hindered production of explosives during the winter of 1914–1915 (see chapter 3), seriously affected

German operational thinking. The general staff could not plan a major offensive and feared that a large Allied push would deplete their reserves of explosives and force Germany out of the war. Furthermore, after the failure of the Schlieffen Plan, the Germans decided against depending on a single measure, in this case Bosch's possible success in producing nitrates from Haber's process, which they did not trust anyway.

The Decision to Use Gas

General Erich von Falkenhayn, who became the chief of the general staff after the German failure at the Marne, appointed a capable officer named Major Max Bauer as liaison between the general staff and the chemical industry. Bauer conferred with the industrialists and determined that some form of asphyxiating gas could be used to overwhelm Allied positions on the front. There were several candidates for this role, all of them by-products of the chemical industry. A bromide was chosen for the first experiment, which was to be conducted on the eastern front in January 1915, although the arguments about the legality of the use of gases were not yet resolved. The experiment was a total flop. In the intense cold of the Russian winter, the gas liquefied and sank to the ground. The military was disappointed and became further skeptical of the scientists. Haber then stepped in and suggested that chlorine would be a better candidate, particularly because it was more abundant. Apart from the technical problems of using a novel weapon on the battlefield, as exemplified by the eastern front experiment, ethical and legal questions persisted.

First, the Hague Protocols outlawed expanding bullets (the so-called Dum-Dum bullets), saw-toothed bayonets, and the use of asphyxiating gases. The German military proved willing to use gas on the Russian front (ironically, Tsar Nicholas II had initiated the Hague Conferences), but using this mode of warfare against the western Allies was another matter. Finally, frustrated by the failures on the western front, the Germans threw caution to the wind. Because of the general adherence to the Hague Convention, such a first use would catch the enemy by surprise and would thus be more effective. Technically, the Germans were not in violation of the convention since they planned to use stationary means of delivery. The Hague Convention forbade asphyxiating gases delivered by "projectiles," but this was generally interpreted more broadly as a total ban.[2]

Implementation

A tremendous logistic and engineering effort was initiated and some six thousand metal containers of chlorine were brought to the front at the Ypres salient. The containers were dug in and covered against possible hits by artillery shells. All this work was performed by the infantry, who carried it out at night so that the preparations would not be observed. Finally everything was ready except that the wind was blowing from the wrong quarter, so there was nothing to do but wait. It is an old truism that soldiers should not be left to idle, and in any case the regular routine of the front line had to be kept going, which meant patrolling, ambushes, and similar activities. In one of these forays, the French captured one of the German infantrymen who actually was employed in the positioning of the gas canisters. The fellow was apparently quite observant, and very talkative, and he spilled the whole story to the French interrogators. To the French the whole episode appeared to be too good to be true. Although there were rumors of German measures for the use of gas (one cannot completely hide industrial preparations on this scale, and it is possible that some information came from the Russian side), the French decided that the "prisoner" was actually a deliberate plant, intended to convey disinformation. It is likely that part of the French reasoning was that the Germans were not so stupid as to send such a well-informed soldier on a mission where he could be taken prisoner.

Action and Failure

Finally, on April 22, 1915, the wind changed direction and the valves on the gas containers were cracked. A low-lying green cloud moved toward the French positions, which were manned by newly arrived colonial troops. The surprise was total and the effect overwhelming. Within minutes some fifteen thousand French soldiers were incapacitated and about five thousand of these were dead. The line of the French trenches was breached on a front of about seven kilometers. More important, there were no real defenses at the rear of this line, and the road to the channel ports, an important strategic objective, was in effect open. The German infantry opposite the affected sector charged behind the dissipating gas cloud, advanced several hundred meters, and then stopped, waiting for the reserves to come through. The German "reserves," however, consisted of one infantry company instead of the three or more divisions that were necessary to exploit this success. Haber, who was sure that the gas weapon would be effective, pleaded with the military to bring up more troops, but to no avail.

After the eastern front fiasco, the German general staff considered the action at Ypres as another experiment and did not want to commit more troops to a dubious venture. This was particularly odd in view of special precautions (which, after all, were not so successful) to keep the whole enterprise a secret. The whole logic of the German general staff was completely flawed. If the weapon was to be a secret, suddenly unleashed on the enemy, why advertise it by a futile and pointless demonstration, even if it was called an experiment? It seems more likely that after the failure to exploit the opportunity presented by the gas attacks, someone invented this excuse to get himself off the hook. Whatever the reasons, it is obvious that certain military minds work in unfathomable ways and a potentially war-winning weapon was wasted in one of the greatest blunders of that war, which, without doubt, boasted quite a few such blunders.

Although badly surprised (at least on the local level), the Allies reacted quickly. Within hours neighboring troops blocked the meager gains of the Germans. One day after the first gas attack, the British Medical Service rushed quantities of sodium bicarbonate to the trenches, instructing the soldiers to soak handkerchiefs in it and tie them around their noses and mouths. Concurrently, an effort was made to develop and distribute an effective substitute for the handkerchiefs. The British army was ready for the second gas attack, on May 1, and by the fall of that year the familiar gas mask, with the canister of chemicals, was developed and distributed. Of course, they were enormously aided by the lackadaisical manner in which the German army initially treated gas warfare. What's more, the Allies, who had previously considered the possibilities of gas warfare (but refrained from any action due to the Hague Convention), now felt free to pursue gas warfare too. Shortly thereafter they actually improved on the German model by devising a gas-carrying shell, making gas warfare independent of wind direction.

EARLY BRITISH TANK DEVELOPMENT

Introduction

Lest it be concluded that the German general staff was unique in this kind of ineptitude in their chosen profession, the opposite side of the front line in that war should also be examined. It is an old maxim that similar problems bring about similar solutions. In addition, before any such solution is accepted—by anybody, but particularly in the armed forces—it is also apt to run into similar kinds of obtuseness. As mentioned earlier, nobody has the corner on sheer stupidity and the Germans were no exception.

Like the Germans, the Allies were stumped by the problem of the stalemate in the trench war, but at least two people on the British side came up with similar solutions.[3] The first was Winston Churchill, then the first lord of the Admiralty. While at first glance it may appear rather strange that the Royal Navy's leader would get involved in problems of land warfare, there were two reasons for Churchill's doing so. For one, the historic mission of the Royal Navy was to protect the Home Islands. Consequently, at the beginning of the war, when it was feared that Britain might be bombed from the air (which eventually did happen), the British Admiralty was asked to look into the matter. Several squadrons of fighters were thus stationed in France and Belgium for that purpose. In order to protect their airfields from a land attack (before the trench lines had solidified into an impenetrable obstacle) the Admiralty established at these airfields several squadrons of armored cars, equipped with rotating turrets mounting machine guns. The crews of these cars occasionally deviated from their assigned missions and made independent forays into German territory. The Germans foiled the British efforts by digging ditches across the roads. Eventually, when the line of trenches got to the sea these raids stopped altogether.

Churchill understood this problem quite well. He had combat experience as a cavalry subaltern in the Battle of Omdurman (Sudan, 1898) at the close of the Mahdist uprising, where Maxim machine guns had been the decisive factor.[4] In 1900 he served as a reporter in Africa during the Second Boer War, but took active part in it, was taken prisoner, and escaped. During the fighting in Antwerp in October 1914, he organized the defense that bought the Belgians five days to get out of a German encirclement. Later still, when he was forced to resign from his position as first lord, he went to France and served in the infantry until June 1916, when he returned to Parliament.

Heavy-caliber artillery pieces (fifteen-inch howitzers), which Churchill had the navy adapt from sea to land use, were too heavy for horses to move around and were dragged in the mud on the front by tractors equipped with endless tracks. Churchill recounts that the master general of ordnance was having doubts about the use of those fifteen-inch howitzers. Churchill relates that the sight of these tractors moving freely over the broken ground was very suggestive. He inquired of the manufacturer if such a tractor could be mounted with armor and guns and launched against the German trenches, and received a positive reply. Churchill then had a suitable design prepared and ordered thirty units. In May 1915 the first of these was tested by the War Office and found unsuitable. The official reasons for the failure were essentially technical. The vehicle could not cross a water barrier of three-feet depth, and it could not descend a step of four feet. While these might have been reasonable requirements, they were somewhat suspect in that the War Office prepared them after the vehicles were ordered. These requirements were not met by any tank until the end of the war.

The order for the other twenty-nine was canceled, but at this stage Churchill did not mind since in the meantime a better suggestion was proposed from another quarter.

Further Developments

Colonel Ernest Swinton was a British officer seconded as an official observer to the headquarters of the British Expeditionary Force (BEF) in France, and whose duty was to report to the War Office in England his impressions of the fighting. He too proposed that a caterpillar tractor be equipped with armor and guns and used to overcome defenses. He wrote several memorandums to the War Office, but the War Office simply was not interested. The true and tested way to overcome defenses was by frontal assaults and not by some mechanical contrivance. Luckily, one of these memorandums found its way to Churchill, who on January 5, 1915, forwarded it with a glowing recommendation to Herbert Asquith, then the prime minister. Another interesting aspect of Churchill's letter was the fact that he proposed establishing a permanent body: "A committee of engineer officers and other experts ought to be sitting continually at the War Office to formulate schemes and examine suggestions, and I would repeat that it is not possible in most cases to have lengthy experiments beforehand. If the devices are to be ready by the time they are required it is indispensable that manufacture should proceed simultaneously with experiment. The worst that can happen is that a comparatively small sum of money is wasted."[5] In 1915 such arguments bordered on prophesy, anticipating problems in R&D decades later.[6] Also interesting is his emphasis on the "experts."

The prime minister forwarded Churchill's letter to the War Minister Horatio Herbert Kitchener, who sent a recommendation to the department of the master general of ordnance, which was part of the War Office. At the end of February 1916, after seven weeks of pondering the problem, that worthy officer wrote a detailed reply, which stated that the idea would not lead to success on account of:

1. The great weights involved
2. The time it would take to design and make sufficient numbers of the machines suggested
3. The vulnerability to gunfire
4. The difficulty of movement over the ground likely to be occupied by the enemy[7]

The master general of ordnance further admitted that there might be somebody who could solve these problems but that he personally was not familiar with such a person. In his memoirs, Churchill was quite blunt in ascribing the difficulties of the development to mechanical defects and official obstruction.[8]

Churchill did not give up. He went back to his armored car officers, who were idle, and urged them to submit a proposal of their own for a vehicle that would be able to cross barbed wire and trenches. These officers rose to the occasion and proposed a vehicle with wheels forty feet in diameter. Churchill had the Admiralty create a committee that analyzed two proposals. One consisted of the vehicle with the forty-foot wheels, and the second of a vehicle with caterpillar tracks. Churchill did not wait for the committee to finish its deliberations but allocated, on his own responsibility, some seventy thousand pounds (approximately $350,000 in 1914 U.S. dollars) to construct six wheeled models and twelve tracked ones, and the contract was awarded to the Admiralty's department of ship construction. Churchill made these decisions on his own because he knew that if he consulted anybody in the War Office or the Treasury, he would have been turned down, and in the event of failure he would have had to answer to a committee of inquiry concerning waste of public funds.

A Qualified Success

In the meantime, however, the Gallipoli landing failed, and Churchill, who was one of its proponents, was forced to resign his position as first lord.[9] His opponents were quick in trying to abolish the tank project on the grounds that this was not the Admiralty's business. But the head of the shipbuilding department (who constructed the vehicle) informed Churchill of this move. Churchill approached Arthur Balfour, who had been nominated as first lord and convinced him to continue the project. Balfour approved the construction of a single-tracked prototype, which was demonstrated in January 1916 and created a very good impression. This was the "mother tank." Construction of fifty tanks was authorized, and they arrived in France in August of 1916. General Douglas Haig, at the time the commander of the British Expeditionary Force, wanted to introduce them immediately to possibly remedy the disastrous results of the opening of the Battle of the Somme. Swinton, who followed the project, pleaded for more time to better prepare the tanks and the crews but was overruled, and the tanks were sent into action. Lack of sufficient training immediately manifested itself. Some tanks bogged down and some crews lost their way and fired on friendly infantry, but about half of them managed to continue until they (literally) ran out of gas. Their advance, however, was more rapid than planned, and there were no ready reserves to exploit the success (a familiar problem). The crews disabled the tanks as best they could and returned on foot to their own lines.

The Germans, who captured the disabled tanks, partially understood their potential, but because of the failure of the British attack came to the conclusion that while tanks could be dangerous, they could not really win wars. It is interesting to note that while the British tanks were going through their difficult development period, Churchill was constantly afraid that the Germans might stumble onto a similar idea and beat the British to it. The Germans started developing a tank of their own but put more emphasis on antitank weapons, heavy thirteen-millimeter rifles and mines. The British tank force finally proved its worth in the Battle of Cambrai (November–December 1917), when Haig used 324 tanks to punch a hole ten kilometers wide in the German line. Again the success was not properly exploited, but this time the tanks' part in the initial success was fully acknowledged, and during 1918 they were used more and more.

Finally Churchill gave credit to Swinton and others, and later admitted that many people contributed to the idea but did not have the executive authority, were not able to command the resources necessary for action, or convince those who had the power to act. He also pointed out that the tank was actually a collection of old ideas, including bulletproof armor, the internal combustion engine, the Pedrail and Caterpillar systems, and of course the armament. He even mentions that the noted author "H. G. Wells, in an article written in 1903, had practically exhausted the possibilities of imagination in this sphere."[10]

DROP TANKS: THE GERMAN FAILURE

Some Technical Points

Fuel is the lifeblood of all mechanized operations, and often operations were mounted or cancelled according to its availability. German Field Marshal Erwin Rommel's difficulties in this respect during the fighting in the Western Desert are quite instructive. It may be embarrassing or even troublesome for a car or tank to run out of gas when under way, but for an airplane it is fatal. Consequently, air operations are planned almost to the minute, with the deciding factor always being the amount of fuel available on board. Naturally, designers strive to increase the amount of fuel to be carried by an airplane, but there are practical and operational limits to this wish. For one, the total payload of the machine is usually of given size, and this has to be apportioned between the weight of the "real" payload, ordnance, or cargo, and the weight of the fuel. While there is a certain amount of trade-off between the two, there are certain lower limits on cargo or ordnance carried below which there is no point in going on the mission. In the twenties, during the "Golden Age of Aviation," stunt pilots carried out refueling in the air, and some remarkable endurance flights were achieved by this method. However, this technology, which today is an everyday occurrence, did not really mature until the jet age, which made it a necessity (because of the prodigious fuel consumption of jet engines) and by disposing of the propeller, made it much safer and simpler as well.

Compared with cargo aircraft or bombers, fighter aircraft are limited in the amount of useful weight they can carry. Within certain limits, the bigger the airplane, the more payload (including fuel) it can carry.[11] But size puts constraints on speed and maneuverability, so it was generally accepted that bombers are big and fly relatively slow but far while carrying a lot of bombs, and fighters are fast and nimble but have a limited radius of operation. A substantial amount of material has been published on the inability of the United States Army Air Force to give fighter protection to deep-penetration bombers (mostly the B-17s) and the consequent losses of the bombers at the hands of the German fighters until the improved P-47 Thunderbolts and P-51 Mustangs came along. Rather less is known of glaring and critical German failures in this field.

During the Battle of Britain, there were reported cases of German fighters turning back because of low fuel and leaving the bombers, which they were supposed to protect, on their own—in effect at the mercy of British fighters. Air combat is usually fought with full throttles, which nearly doubles the fuel consumption, compared with economic cruise flight. Furthermore, after raids over Britain, the Germans still had a long flight home, while the British fought literally above their own air bases, giving them a considerable advantage, fuelwise. Cajus Bekker quotes a German fighter pilot who claimed, "[T]here were only a few of us who had not yet had to ditch in the Channel with a shot up aircraft or stationary airscrew."[12] This situation applied to operations in the south of England, but what about action farther afield?

"Black Thursday" of the Luftwaffe

On August 15, 1940 (during the Battle of Britain), the Germans staged an attack from Norway and Denmark across the North Sea against targets on the eastern coast of central England. The distance was more than four hundred miles, farther than possible for any single-engine fighter cover (the Me-109 had at that time a radius of action of less than two hundred miles). The bombers were accompanied by twin-engine fighters (the Me-110), but it was widely acknowledged that these aircraft were no match for single-engine fighters. Before and during the initial fighting of August, the Germans misled themselves with their pilots' glowing but erroneous reports of "kills" of British fighters. The Germans also assumed that the heavy fighting in the south would draw away enough of the RAF's fighters so that the bombers on this northern raid would have more or less a free run. This hope was dashed when the attacking force was pounced on by completely unexpected British fighters and badly mauled. Bekker reports that the Me-110s had external, jettisonable fuel tanks, what are today referred to as "drop tanks" or auxiliary tanks.[13] By the time the Germans reached England, the auxiliary fuel tanks were almost empty and thus filled with a mixture of gas and air, but because of a mechanical malfunction, the pilot of the flight commander's airplane did not succeed in discarding it. Apparently, during the engagement, the empty fuel tank was hit and exploded. Bekker further comments that during the fighting in Norway (in early 1940), many pilots were lost because of a similar malfunction.

Drop tanks were developed toward the end of World War I by the Americans and French as a safety feature to be used in case of fire. After the war, they continued to be tested and used for range extension. On most American aircraft the attachment points for the external tanks also served for bombs. As aircraft technology improved, aircraft were designed with larger internal fuel capacity. When that evolved, the army issued a prohibition on external racks for fuel tanks—partly to lessen the additional drag but mostly out of fear that ground commanders would use the fighters to attack ground targets by bombs hung from the fuel tank racks.[14] The navy did not follow this course, and its aircraft could carry, externally, either fuel or bombs.

The Germans apparently followed these developments. Drop tanks were first used operationally by the Condor Legion (a force of German "volunteers" who fought on the side of Franco in the Spanish Civil War, July 1936–March 1939) to extend the range of fighter aircraft. In Spain it became obvious to the Germans that bombers could not sortie without fighter protection. Adolf Galland, a noted German fighter pilot and later inspector general of fighters, complained that drop tanks were not used by German fighters during the Battle of Britain.[15] From the above report it seems that they were used to some extent, and apparently the crews eyed them with suspicion because of some fault in their release mechanism. The facts are thus indisputable. The Germans pioneered the combat use of drop tanks, used them to a limited extent on operations, but failed to adapt them for use by single-engine fighters. This omission limited the combat range of the single-engine fighters, which could not operate farther north than London, and even this range was barely sufficient and resulted in losses not related directly to combat.

The range limit had another crucial outcome. The Germans learned quickly that they could not send unescorted bombers up north, and thus that area became a haven, both for British war industry and for fighter squadrons that were being rotated there after the intense operations in the south. This became evident and was commented upon practically from the start of the fighting.[16] There the crews could rest and train new pilots, and the northern squadrons became the reserve squadrons for the big engagements, while effectively being beyond the reach of the German bombers.

So the obvious question is, Why didn't the Germans fix the problem of the faulty release mechanism if it really was the problem? At least two other contemporary fighters had drop tanks. The first was the French Morane-Saulnier-406, which first flew in 1935 and fell into German hands when France was occupied. In the summer of 1941 aircraft of this type were ferried to the Middle East to support the Vichy forces in Syria and Lebanon against the British attacking from Palestine. While this particular operation took place after the Battle of Britain, it can be surmised that these aircraft were initially designed with drop tank capability, since the French had to worry about their empire, from North Africa to the Middle East. The second contemporary fighter with a drop tank was the Japanese Zero, which first flew in April 1939. In any case, there is no question that the problem had a solution. In fact the Germans themselves used drop tanks later in the war, and the Americans never had such a problem. As mentioned above, the problem first appeared during the Norway campaign that took place between April and June 1940. That left the Germans some three months to rectify an essentially mechanical fault, which was perfectly within their engineering capabilities. In addition, as we shall see in the story of the upward firing gun (see chapter 5), the Germans had a somewhat cavalier, but quite practical attitude toward field improvisations, which even at that time would have caused any Allied technical officer to tear out his hair.

Drop Tanks and the Me-110

So what caused this German failure? One possible conclusion is that they decided to trust what they already had in hand, including the capabilities of the Me-110 as a fighter and the range of their single-engine fighters. But when discussing the Me-110, another puzzling question arises. Williamson Murray writes that the Germans did not bother about drop tanks "most probably because they believed that the Bf-110 [a different designation for the same airplane] could successfully perform the mission. When they discovered that it could not, it was too late."[17] This comment reflects on the Germans in an even worse way. There is no question that German aircraft designers were quite a capable lot. Willi Messerschmitt, Kurt Tank (of Focke-Wulf 190 fame), and Ernst Heinkel could hold their own against anybody in the Allied camp. So they—particularly Messerschmitt,who had designed the Me-110—understood the limitations of twin-engine fighters in combat against single-engine ones, and they knew the capabilities of the Spitfire and the Hurricane since several of these had been captured in France. Did the Germans never test the Me-110 in mock battle against one of their own single-engine fighters, or even better, against a captured British one? If this was never done, then the German aviation industry and the technical and intelligence branches of the Luftwaffe stand accused of neglecting a basic procedure and a gross intelligence failure. If such an evaluation and mock combat had been performed, the results would have been obvious, and in this case the German high command is to be blamed for an even worse blunder or for totally unrealistic expectations that events would turn out differently in the skies over Britain. But one does not enter a modern war on the strength of a belief or expectation. That is tantamount to trusting divine intervention to happen at the right moment. It might well happen, but it is hard to fit into an operational plan.

This attitude, however, ties in with Alexander de Seversky's comments about his visits to the German aircraft industry.[18] De Seversky claimed that the German aviation industry was producing sleek, technically advanced aircraft, but their overall suitability for combat did not progress past World War I thinking. All these aircraft had were numbers, but their inferiority in other aspects was so pronounced that mere numbers could not compensate for it.

One other possible explanation is that the vaunted German order and organization was nothing more than a myth. It seems that these paragons of precision and detail were nothing of the sort and that the basic requirements of documentation, systemwide collation of data, and "institutional memory" were lacking in Germany during World War II. This also ties in, partially at least, with the German failure to appreciate radar in the early warning role.

THE MISSED OPPORTUNITY OF THE SNORKEL

All of the operational submarines of World War I and all those of World War II were equipped with a dual propulsion system: a set of diesel engines for propulsion when the submarine travels on the surface, and electric motors, drawing current from batteries, for submerged operations.[19] When on the surface, one of the diesels was usually used to charge the battery system to full capacity. This arrangement was necessary because the diesels need air for combustion, like any truck engine, while electric battery capacity, even for battery packs weighing tons, is rather limited. The nuclear power plant changed all this, and nuclear submarines can travel underwater for years without refueling. The amount of oxygen required for breathing by the crew is minute (in comparison with even a modest internal combustion engine), and various chemical systems suffice to keep the atmosphere inside a nuclear submarine quite fresh. It should be pointed out, however, that even today only the big powers own nuclear submarines, and the rest of the world still makes do with the diesel-electric submarines. Due to the limited capacity of the batteries, and the fact that on the surface a submarine is relatively vulnerable, the idea of a submarine operating continuously under the surface while powered by other means than electric motors was quite enticing.[20]

Before World War II, the German navy discussed the possibility of fitting its submarines with a snorkel, an extendable pipe that would draw air while the submarine was at periscope depth.[21] However, apart from discussing the idea the German navy did nothing about developing such a useful feature. I. I. Wichers,[22] a Royal Netherlands Navy commander, is credited with inventing the snorkel and having it installed in 1937 in Dutch submarines of the O-21 class.[23] He was not the first, though. On September 12, 1800, in one of his experiments in France, Robert Fulton used such a snorkel in the sea off Normandy, and his submarine stayed underwater for six hours with no ill effects for the crew.[24] The snorkel consisted of a pipe or tube, extending about 7.5 meters above the top of the conning tower. This pipe had two pathways—one for the incoming air and one for the exhaust gases—and a valve that closed automatically if a wave got to the top of the pipe. A submarine thus equipped could run for extended periods underwater, while keeping its batteries fully charged for a "real" dive. Because of its small size, it was practically impossible to visually spot the pipe's head above the water, unless the sea surface was dead calm. (It should be remembered that we are talking about visual detection. Radar was not yet developed enough for that purpose.) When Germany invaded the Low Countries in 1940, one of the snorkel-equipped submarines (O-21–24) escaped to England. The Royal Navy examined the innovation, decided it was not needed, and had it removed.[25] Two other such submarines fell into German hands.

When the dust settled after the campaign in the Low Countries, the Germans examined the Dutch installation and came to the conclusion that while the snorkel was ingenious in its design, it was of no use to the German submarine arm. For one, the war was practically over, or at least that was the prevalent opinion. Second, on the drawing boards already was an advanced submarine with a new type of propulsion that did not require atmospheric oxygen for its operation.[26] These new submarines were named the Walther submarines for the developer of the novel propulsion system. The new system was based on technology that was originally developed for torpedo propulsion and was based on the decomposition of highly concentrated hydrogen peroxide—H_2O_2. The introduction of the Walther submarine would make the snorkel redundant, so it was shelved for the duration.

Hydrogen peroxide is a nasty material, and its introduction as a fuel suffered from many teething problems.[27] The system worked well enough in the laboratory, but when scaled up and put into the environment of an operational submarine, there were problems. Besides, for a long time the chemical's development as a fuel was slowed by the ban on long-range projects that Hitler had instituted after Germany's first round of victories. When in 1943 the German submarine arm felt the noose tightening, they quickly introduced, as an interim measure, the new Type XXI submarines, which were much bigger, with greater battery capacity, and hydrodynamically, a much improved hull. It should be remembered that until then submarines of all nations were basically surface crafts that could also dive. They were constructed with flat decks, guardrails, deck guns, and many other protuberances, all of which created drag when moving underwater.[28] All these drag-producing appendages were eliminated, and the new submarines were capable of an underwater speed of eighteen knots, compared with about four knots for the standard submarines. They were also equipped with the snorkels, which made them much more difficult to detect.

Luckily for the Allies, at about that time a new, higher-frequency radar was perfected. This was the three-centimeter (10,000 MHz or 10 GHz) radar, which was capable of detecting targets as small as the snorkel's head. The first radars of this kind were mounted on Mosquito aircraft that were equipped with fifty-seven-millimeter guns. The big advantage of the Mosquito was that it was quite capable as a fighter. It could fight or, when necessary, evade German fighters, unlike the obsolete light bombers and even the lumbering flying boats that were used by Coastal Command. Furthermore, it did not have to overfly the submarine in order to attack it and thus was less endangered by the submarine's antiair armament.

After the war, the British tested their older ten-centimeter radars against captured German submarines equipped with snorkels, with most enlightening results. The radars managed to affect detection in only about 6 percent of the tests. In other words, if the Germans had installed snorkels on their submarines immediately after they discovered the device in Holland in 1940, or even as late as 1941, when they captured the first airborne antisubmarine radar, they might have acquired a tremendous advantage.[29] Furthermore, at that stage, the Germans should have foreseen the possibilities of disaster and made an all-out effort to improve their submarines. If this had happened, it is quite possible that the Battle of the Atlantic would have ended (or at least continued) differently. It is possible that the submarine could have continued successful operations for a longer time before being checked by the various Allied developments. Even the increased production of cargo ships in American yards might have proved to be insufficient, and it is possible that the buildup of resources for the invasion would have been slowed. It is unnecessary to speculate further on this point, but the argument is quite clear: the Germans committed a gross error of judgment on the desirability of introducing a technological innovation.

While the British rejection of the snorkel was foolish, it was harmless compared with the German blunder. In contrast to the Germans, the British did not base their naval strategy on submarine warfare, and thus the adoption of the snorkel, though it might have proved useful, was not critical. For the Germans, however, this mistake may have been fatal.

This whole affair is very much reminiscent of the drop tank episode described above and strengthens a previously reached conclusion. The Germans, although the aggressor in both world wars, always thought of war in terms of very brief duration. However, their overwhelming successes at the beginning of the war were due as much to poor Allied planning and sluggish reaction as they were to German design. While the Germans introduced many innovative systems and technologies, many of these came too late in the war. This also points to a recurring tendency to neglect thinking things through, in an orderly manner, to a logical outcome. This is doubly strange, as the Germans are generally considered very methodical and precise in their thinking.

THE PROBLEM OF THE ALLIED LONG-RANGE ESCORT FIGHTER

The fact that B-17s continued to be sent on deep-penetration raids without long-range fighter escort should not be construed as meaning that their commanders were unfeeling old men who did not care about losses. But within the constraints of having to carry out their mission, the commanders became captive to two assumptions from which they had difficulty extricating themselves, even when both these assumptions were proved totally wrong. The first assumption was that the heavy bombers would be able to take care of themselves. As we have seen, this was perfectly true when the mission—strategic bombing—was conceived. Of course, it can be asked why nobody saw the technological indications and changed the bombers' armament specifications, or for that matter changed the concept to include the long-range fighters in the equation. Furthermore, it appears that the thinking on this subject was somewhat confused. In a report of the air corps from May 1939, it was stated that "the higher operating speed of modern bombers increases the difficulty of interception by hostile pursuit aircraft and thereby lessens the need of support by friendly pursuit."[30] Later in the same report it was asserted that "American pursuit should be designed primarily for successful interception and destruction of hostile bombardment over or near friendly territory." Well, if American pursuit could do this, why couldn't the hostiles? But the second assumption is even more puzzling.

The question of escorting long-range pursuit was raised several times and opinions were divided, but in the end it was decided, essentially by default, that the bombers would have to go it alone. From the end of the First World War, it was thought impossible to build a good fighter with enough range to accompany multiengine bombers. Experts said "it could not be done" and it was assumed that this was so. But the truth is that nobody took the time to check the figures. To start with, the Americans thought the idea was superfluous because in the United States nobody thought in terms of long-range fighting. The position of the army was that airpower is important only to a distance of two hundred miles, to protect the ground forces from the enemy's air units.[31] Finally, as pointed out in the previous section on the German drop tanks, the U.S. Army was opposed to drop tanks. The intellectual environment in England was similar. In March 1940, before the Battle of Britain, Air Marshal Hugh Dowding, commander of Fighter Command, wanted to have a long-range fighter designed. He was curtly told by a committee from the Air Ministry that it was impossible to achieve such a fighter with a performance comparable with a regular short-range fighter. Apparently, the committee was thinking in terms of a twin-engine aircraft and was probably right about the limitations of such an airplane in a dogfight. In this respect, at least, the committee did its work more thoroughly than the Germans with the Me-110. A year later, Churchill received the same answer from Charles Portal, chief of the air staff.[32] The supposed impossibility of a long-range escort fighter was the accepted wisdom of the time and no one bothered to pursue the matter further. This attitude was plain intellectual laziness. If someone had bothered to do a preliminary design for such an airplane, the notion of the impossibility of such an airplane would have been immediately shattered.

A few months later the imagined aircraft became a reality with the design of the P-51. The fact is that the technology (not the design) of the Mustang (P-51) and the Thunderbolt (P-47) was not that much advanced, compared with that of all other contemporary fighters. The P-51 was conceived in 1940 and flew in 1941. The P-47 was designed at the end of the thirties and first flew in 1941 as well. For the P-47 the solution consisted of an aerodynamically "cleaner" design and a huge internal fuel capacity. Admittedly, the P-47 could not hold its own in a real dogfight against the more nimble short-range fighters. So a technique was adopted that had been developed by General Claire Chennault, the founder of the AVG (American Volunteer Group—the Flying Tigers) under similar circumstances in China. The P-47s would attain altitude and dive through the enemy formations. For the P-51 the solution was somewhat more ingenious and built into the aircraft from the start. The P-51 was designed with a low-drag airfoil, the "laminar airfoil," which although it had other problems was ideal for long-range flights. Admittedly, the value of this part of the improvement is debatable. It was later found that the effectiveness of the laminar airfoil depended on the pristine condition of the wing's surfaces, which was hard to maintain in squadron service. Additionally, the oil-cooling system of the P-51 was designed in such a way that when the heated air escaped, it provided some thrust, like a miniature jet engine, and enough to cancel the drag of the cooling system. But then the P-51 ran into a snag. When the British ordered the airplane from North American, it was delivered with an Allison engine, which was somewhat "anemic" above twenty thousand feet. The British then put in the "Merlin" engine (the power plant of the Spitfire) that completely changed it. The Army Air Force received two Mustangs from North American but was not impressed. Murray writes that "the road to production, however, was not easy; there was reluctance to push its development, since it was not entirely a home-grown product."[33] The American assistant air attaché in London put it more graphically: "Its development and use in this theater has suffered for many reasons. Sired by the English out of an American mother, the Mustang has no parent in the Army Air Corps or at Wright Field to appreciate and push its good points."[34] This was a clear case of NIH (Not Invented Here), and it ties in well with the affair of the Sherman tank and the British seventeen-pound gun described below. Only in the spring of 1944 did P-51s start arriving in England in meaningful numbers to supplement the P-47s.

But as previously shown, the real solution was more fuel, and since internal volume in fighter aircraft was limited, the added fuel had to be carried in external (drop) tanks. The tanks were invented in the United States originally as a safety measure against fire, and by 1940 they were a familiar technology worldwide.[35] The Germans and the French used them, and on the other side of the world the Japanese used drop tank–equipped Zeros in China. General Chennault, who was then in China, reported this, but his report was ignored. Zeros equipped with drop tanks also participated in the attack on Pearl Harbor.

In the spring of 1942, British Hurricanes (Mark IIC) equipped with two forty-four-gallon drop tanks, started making fighter sweeps over occupied France. At cruise power these aircraft had a range of 920 miles. This was not yet enough to get to Berlin and back. The Eighth Air Force apparently knew about these developments, because in October 1942 it started requesting drop tanks for the P-47s, but nothing happened. It is possible that until then there were those who still believed they could continue bombing operations without fighter escort and were trying to prove the point. With the growing bomber losses over Germany, pressure started building up and the Eighth Air Force in Britain contracted for some British-made fuel tanks as an interim solution for its fighters. It appears that in the United States there was administrative confusion about the question of drop tank production,[36] and "combat tank production and development continued haphazardly."[37] The disaster of Schweinfurt on October 14 finally galvanized everybody into action, and in November 1943 sufficient numbers of drop tanks started arriving in the squadrons in England, which spelled the beginning of the end for the Luftwaffe. One wonders how the air war would have looked if both the P-51s and suitable drop tanks had been introduced in late 1942, or early 1943, when technically it was already possible to do so. It is interesting to note that even before D-Day, three British Spitfires, a notoriously short-ranged airplane, were sent to the United States, equipped with drop tanks, and returned, flying, to England.[38]

PGMs (PRECISION-GUIDED MUNITIONS) IN VIETNAM

Introduction

PGMs are defined as weapons that can be controlled after they are shot or launched, so as to give them a higher probability of hitting their target. This can be done either automatically, by some kind of guiding device in the weapon, or under the control of an operator, who has information about the relative positions of the weapon and the target and can guide the weapon by remote control. The earliest such device was probably the "Kettering Bug," developed during World War I in the United States. It was a small pilotless plane, gyroscopically controlled, which was loaded with three hundred pounds of explosives and designed to fly toward a target and explode. It could fly to a distance of up to a mile with reasonable accuracy and in essence was the first cruise missile. The war ended before the plane entered use.

In World War II both the Germans and the Americans used remotely controlled missiles and bombs. In the Mediterranean and the Bay of Biscay, the Germans scored several hits against Allied shipping with both radio-controlled and wire-controlled (conducting electrical signals) weapons, both glide bombs and powered missiles. The Americans developed a radar-guided weapon, the Bat, and used it successfully against Japanese sea targets. Television and heat-seeking guidance systems were also developed. The most important weapon developed, which also saw extensive operational service, was the AZON, a standard bomb with a radio-controlled guidance kit attached to it. The AZON was successfully used against small important targets both in Europe and in the Far East, scoring hits against targets that had previously survived numerous attacks by conventional means.[39] AZON is the acronym for AZimuth ONly, meaning that the bomb was steered only in direction. Toward the end of the war an improved version, RAZON (Range + AZON) was developed, which could also be steered in elevation. The one thousand–pound RAZONs were used in Korea but initially, because of long storage and lack of crew proficiency, did rather poorly. With time, results improved and a much larger bomb (twelve thousand pounds) was adapted for this mode of use. Although the bomb was successfully used, the Far Eastern Air Force Bomber Command recommended its withdrawal, citing lack of targets and danger to pilots. The U.S. Air Force Official History, on the other hand, praised the successful results and cited the RAZON project officer, who claimed he could "prove that one guided bomb is worth one thousand conventional bombs against 'line targets'—bridges, etc."[40]

The Use of PGMs in Vietnam

After the Korean War, development of weapons for conventional war stagnated, with emphasis and funding going mostly to preparations for massive nuclear war. The Vietnam War renewed interest in conventional weapons, mostly because it became obvious that this war required weapons that were different from those intended for the European battlefield. The air force made some organizational changes, and a small research group, tasked with improving response to tactical operational needs, was established in 1964.

On the advice of civilian researchers, laser technology was considered in order to improve the accuracy of free-falling bombs. This led to a debate in the air force, pitting improved accuracy against the danger to the loitering plane that had to remain on station to guide the bomb. Other technological solutions (such as TV and infrared) that enabled "launch and leave" were considered, but nevertheless the laser system received the highest priority, and after a series of tests, production of one hundred units per month was started in late 1967.

Air Force General John Lavelle tried to convince his service to adopt "smart bombs" on a large scale, on the order of three thousand per month, but his suggestion was rejected. The reasoning given was that the air force's munitions budget was determined on the basis of bomb tonnage dropped in combat. If smart bombs were used, missions could be accomplished with considerably less tonnage of bombs and the air force's budget would suffer.[41] This answer was given by a general officer, and apparently no consideration was given to such questions as better success rates, lower sortie rates, and inevitable losses stemming from more sorties flown. In 1968, on a trip to Southeast Asia, U.S. Air Force Secretary Harold Brown suggested to the air force commanders that he would try to convince the secretary of defense to increase the production rate of smart bomb kits to six hundred per month. The air force politely declined. This was at a time when the air force was flying twenty thousand fighter-bomber sorties a month. Apparently, by 1971 the commanders of the Seventh Air Force had not yet made up their minds whether or not smart bombs were cost effective.

An official report was very explicit on this point: "The logic was bizarre. The standard argument held that it was wasteful to expend a $3,000 bomb on a $1,000 truck. This analysis ignored the obvious point that spending ten $1,000 bombs on the same target was even worse."[42]

In 1971 General Lavalle (who still believed in smart bombs) became commander of the Seventh Air Force. He immediately requested an increase in the production rate, and obtained an increase to three hundred to four hundred smart bombs a month. In the spring of 1972, with the resumption of air bombardment in Vietnam, these bombs were extensively and successfully used, obliterating some targets that for years previous had eluded destruction and had cost numerous losses as well.

One of the conclusions in the above-quoted document states that "generals and admirals must bridge a 'conceptual gap' between comfortable routines of the past and abruptly new ways of doing things."[43] This apparently is true for generals and admirals the world over. Another important lesson from the above affair is that forceful individuals in the right place can change ingrained attitudes, and in certain organizations this depends on the individuals manning and running the organization rather than on any inherent fault in the "organization's" makeup or mission.

5 Bad Management, Worse Leadership, and NIH

There are over two thousand years of experience to tell us that the only thing harder than getting a new idea into the military mind is to get an old one out.

Liddell Hart

THE CENTIMETRIC WAVELENGTH RADAR BATTLE

Preliminary Skirmishes

Even before the outbreak of World War II, the British were well aware of the German progress in submarine design and the circumvention of the Versailles agreement limitations, although they did not think the Germans would again revert to unrestricted submarine warfare. Ironically, initial funding for continued German submarine development came from the Germans' sale of ships that had to be scrapped under the Versailles treaty. The Germans established a network of plants, some of them outside Germany, that produced components and systems for submarines, and finally complete submarines, and sold them to foreign governments, while gaining valuable experience in modern technologies. Some of these submarines were in fact prototypes for future German submarines. The submarine fleet became public after the Nazis came to power. The British, however, were somewhat complacent because they considered the ASDIC—developed toward the end of the last war but never tested in action—as a potent enough tool to deal with the submarine. Unfortunately, while the ASDIC's range was about a kilometer and a half, current torpedo ranges were from five thousand to fifteen thousand meters, with some experimental types going even farther. Moreover, after the fall of France, the Germans had the use of all the French ports, and in order to sortie into the Atlantic, the German submarines did not have to go through the English Channel or through the North Sea and around England. Consequently, although initially disdaining convoys again, the Admiralty quickly was willing to listen to new ideas.

Fortunately for the British, an outgrowth of the aircraft-detection radar work proved useful in a new role. In order to supplement the radar ground stations, it was thought that an airborne radar would be useful. The problem faced by the British was to reduce the radar system and its antenna to a size that would fit into an airplane. This was successfully accomplished in the summer of 1937 (with a radar wavelength of 1.25 meters), and the development crew went flying to look for other aircraft. They did not find airplanes but found ships at sea, and thus a whole new horizon—for radar research and development—had opened. It also turned out to be quite successful in detecting submarines on the surface, and eventually all of Coastal Command's aircraft on antisubmarine duty were equipped with such radars, the so-called ASV (Air-to-Surface Vessel) Mark II, with a wavelength of about 1.5 meters.

One must remember that the submarines of that period were diesel-propelled submarines with an auxiliary electrical system for underwater operation. The submarines usually operated on the surface by means of their multidiesel engines (one of which was used to charge the lead-acid batteries for the electrical system) and dove only during the day or when under severe attack. Submarines leaving their bases in occupied France, on their way to the Atlantic Ocean, and traveling through the Bay of Biscay, traveled at night on the surface and dove during the day. By using the radars, patrol aircraft were able to detect and attack the submarines on the surface and at least force them to submerge. This strained the crews, forced them to shorten battery-recharging times, and slowed the submarines' transit through the Bay of Biscay.

The Leigh Light

It was quickly discovered that because of spurious signals returned from the sea surface (sea clutter), the minimum range of these radars was about one mile. In other words, after the aircraft discovered the submarine on the surface and started its attack run, when it got to the one-mile range, the target had disappeared from the radar screen. A distance of one mile was equivalent to approximately fifteen to twenty seconds of flight time. On a bright night it did not make much difference, because at that range the crew could acquire the submarine visually, but on a dark night this was hopeless. Coastal Command was becoming desperate and circulated requests for suggestions.[1]

Squadron Leader Humphrey de Verde Leigh had been a fighter pilot in World War I and flew antisubmarine patrols at that time. Now he worked at Coastal Command headquarters. One day a pilot from the command came to straighten out some administrative details, and when during the ensuing talk he found out that Leigh was an ex-colleague, he poured out his frustrations. This indiscretion was of course an infringement of at least half a dozen sections in the Official Secrets Act. Leigh listened, mesmerized—in World War I, of course, there were no radars, efficient or otherwise—and after his new friend left, he kept thinking about the problem. Within three weeks (in October 1940) Leigh prepared and submitted a detailed memorandum in which he offered a solution. He suggested mounting a large, retractable searchlight on the belly of the airplane that would be bore-sighted with the radar's beam. When the radar's image started to disappear, the searchlight would be switched on and the attack run continued. Leigh was a thorough professional. His memorandum included details about installation, aiming, and even ventilation of the big thirty-six-inch searchlight. (This incidentally was later reduced to twenty-four inches.) The memorandum was sent to the various departments, including the RAF's flight-test department in Farnborough. Everybody thought the idea was great, except the scientists at Farnborough, who found it full of faults and suggested that everybody wait until their better version would be shown, sometime in the future. It was a clear-cut case of NIH (Not Invented Here), in which the professionals were incensed by upstart amateurs stepping on their turf. Leigh's superiors, however, supported him, and the first installation of the Leigh Light on an airplane was made in March 1941. It was thoroughly tested, and in the fall it went into series production. In December the first submarine was sunk with the help of the Leigh Light.

The Germans, however, had known about this radar installation since the beginning of 1941, when Rommel captured one in North Africa. When submarines started disappearing in the middle of the night, the Germans put two and two together and came up with the right answer. They installed receivers for the wavelength of the device they found in the captured bomber (176–220 MHz, or the equivalent of a wavelength of 1.36–1.70 meters) in three of their submarines. Soon it was found that these receivers could discover the radar's emanations at a range of some fifty kilometers, way beyond the useful range of the radar itself. The German electronics industry was busy to its full capacity, and the order for the new receivers was given to a French company named Metox. The Germans were quite satisfied with themselves, so much so that Admiral Karl Doenitz, the commander of the German submarine force, claimed that "the airplane would win against the submarine just like the crow can fight the mole."

The Centimetric Wavelength Radar

The electronics industry of the period had a "holy grail" to seek. Everybody was searching for a practical way to produce even higher-frequency radar systems. The sought-after target was the so-called centimetric radar, with an operating frequency of 3,000 MHz and higher, equivalent to a wavelength of ten centimeters or less. The use of these frequencies enabled smaller antennas, better resolution, and a host of other technical improvements. Although suitable devices to produce this range of frequencies were developed by researchers in many countries, they were useless in practice because they could not deliver enough power for a truly useful radar. On the average they delivered up to forty watts. Some of these devices were controlled by a magnetic field and thus were named "magnetron." Furthermore, after some theoretical calculations, German scientists erroneously concluded that even if enough transmission power could be obtained at such a frequency, such a device would not work. At these wavelengths too much of the reflected energy would behave like light and go out in all directions, and too little of it would actually be reflected back toward the point of origin and made use of. And this after all is the essence of radar. It seems that in this case the work was misapplied and the Germans actually discovered a solution to an as yet nonexisting problem—stealth. Since in their thinking they approached the problem from the wrong direction, their conclusion was of course that centimetric radar does not work. It might be historically interesting to ascertain this point.[2] The debates on this point, within the German scientific community, blossomed into a first-class row, and work on shorter wavelength was essentially stopped or slowed for nearly a year. This situation affected personal relationships between the researchers to such an extent that all work was actually hampered. At that point, January 1943, Major General Wolfgang Martini, the head of the Luftwaffe signal organization, had to step in and formally order a stop to all centimetric wavelengths work because of its low prospects for success.

On the other side of the channel they had never heard of this development. Two British scientists at the University of Birmingham, John Randall and Henry Boot, were given a low-priority assignment to develop a receiver for very short wavelength radiation. To test their receiver they needed a suitable transmitter, and until the receiver was built they looked around for such a transmitter. They combined an American invention, the Klystron, with the old magnetron tubes and came with a device they also named the magnetron. After several tests, in February 1940 it was found that the new device delivered the unheard of power of five hundred watts at a wavelength of 9.5 centimeters. Here Randall and Boot got some help from a French scientist who contributed another bit to the development that doubled the power output. By the time these transmitters went into production, the output power on new radar systems rose to more than ten kilowatts.

The British realized that under the prevailing wartime conditions they did not have the resources to properly exploit this new development. Despite its immense military and commercial value, they added the magnetron to the list of items that the Tizard's scientific mission took with it to the United States in September 1940. The aim of this mission was to exchange scientific information and to explore venues of cooperation. Churchill, however, decided that all the British offerings would be an outright gift, and the mission took with it details of various British developments. These included the cream of the British scientific developments, among them the (Whittle) jet engine, details of the Rolls Royce Merlin engine (the Spitfire's power plant and soon to power the Mustang P-51), data on Britain's work on atomic research, and the magnetron. While all the other information was interesting, the magnetron caused a considerable stir in U.S. scientific circles, which had already tried to develop a useful centimetric radar and failed because of the low power output. The magnetron was eagerly accepted by the Americans, who started a wide-ranging development and production effort in a field that promised great utility. Moreover, the whole effort was in a frequency band that the Germans did not suspect possible and considered science fiction.

The new, magnetron-based radar was first test-flown in March 1941 and, among other capabilities, proved an adequate device for "reading" the ground as a blind bombing aid for bombers. In the meantime, the radars went into production and an argument ensued over who would get these radars first—Bomber Command or Coastal Command. (The night fighters already had them.) The situation escalated into a political fight between the services, this time pitting Lord Cherwell (Frederick Lindemann), who supported Bomber Command, against Tizard and Blackett (a member of Tizard's CSSAD back in 1935), who supported the claims of Coastal Command.[3] Robert Watson-Watt, the British radar pioneer, supported Bomber Command in the debate. Two such sets were on the *Prince of Wales*, which on December 10, 1941 (together with HMS *Repulse*) was sunk by Japanese bombers in shallow waters off the coast of Malaya. Furthermore, the Japanese raised the ship and it was possible that they found the radars and informed the Germans. Finally, Watson-Watt claimed that it was possible the German monitoring service picked up the transmissions from the sets already in use. He predicted it would take the Germans two to three months, after such a radar fell into their hands, to develop listening receivers.[4]

As an aside, it is noteworthy that Bomber Command later refused to transfer bombers to Coastal Command for long-range escort and submarine-hunting missions. The operations research people, of whom Blackett was one, calculated that each long-range bomber stationed in Iceland on antisubmarine patrols would save six ships over its lifetime. The same bomber on bombing missions over the continent would kill no more than two dozen Germans.[5] Richard Overy (quoting a previous source) put this in the proper perspective when he wrote: "Why the RAF remained resistant for so long to the idea of releasing bombers for work over the ocean defies explanation. A mere 37 aircraft succeeded in closing the Atlantic Gap, whose existence had almost brought the Allies' war plans to stalemate."[6]

Toward the end of 1942, Coastal Command received the first forty centimetric radar units, but there were delays in operational use because of problems of installation. The later radars went to Bomber Command and Fighter Command. At the beginning of 1943 the radar (called H2S) was approved for flights over enemy-occupied territory. On the night of February 2 to 3, a bomber carrying the H2S radar was shot down near Rotterdam in Holland. The radar installation was damaged, but the all-important metal magnetron valve survived the crash.

The German engineers were fascinated by the prize that fell from the sky into their laps and the unusual technology—the small antenna, the wave guides, and most important, the multicavity magnetron. They named the device the "Rotterdam." A few days later, the German laboratory investigating the device was destroyed in an air raid, but luckily for them, on the same day another H2S-carrying bomber was shot down and the German investigation was not significantly hampered.

The Effect of the Centimetric Radar on the Germans

When the magnitude of the find finally dawned on the Germans, they were hardly happy with it. When Goering received the "Rotterdam" report, he expressed their feelings: "I expected the British and Americans to be advanced, but frankly I never thought that they would get so far ahead. I did hope that even if we were behind, we could at least be in the same race."[7] The Germans immediately ordered that several copies of the new radar be produced for study. Since they appreciated what this radar might do to the submarine force, they also ordered warning receivers for the new frequency, which were dubbed Naxos. In the meantime, Coastal Command got the radars into operation, and in February the first submarine was attacked without being warned by the old Metox device. The submarine survived and even shot down the attacking airplane, but the Germans realized that changes would need to be made. However, they ran into production difficulties with the Naxos, and when things started going from bad to worse for the submarines, they were ordered to dive at night and surface during the day only for battery charging. Nevertheless, the graphs kept to chart submarine sinkings started climbing again. At the beginning of May, Watson-Watts's three months had passed, but the Germans still could not produce any useful countermeasures.

The German awe at the Allied prowess in electronics brought some additional windfalls for the Allies. The range of the centimetric radar was fairly short, and once in the open sea the submarines were quite immune from being discovered. The Allies had partial information about submarine location from deciphering the Enigma transmissions and from "Huff-Duff" (High-Frequency Direction Finding), the triangulation of transmissions sources. The Germans, who were absolutely sure that the Enigma was secure, used it extensively, sometimes quite frivolously. The large amount of useless talk being transmitted simplified the work of the code breakers but also helped the Huff-Duff listeners. One thing led to another, and a German security officer made another silly mistake that masked even the possibility of finding out about the Huff-Duff. German agents working from Algeciras (a Spanish port near

Gibraltar) photographed Allied ships in port in Gibraltar. On their masts were the typical Huff-Duff antennas that would have disclosed the secret to any first-year student of electronics. When the Germans obtained the pictures, they retouched them in order to hide the background, which could betray the position from where they were taken, and then distributed the pictures to the navy. However, in the process of retouching, the special Huff-Duff antennas were also deleted from the photographs.

The Allies followed the iron rule that no operation should be initiated only on the basis of an Enigma decrypt, so as not to disclose it as a sole source of the intelligence. But the volume of radio traffic gave the Allied intelligence enough initial data to send a reconnaissance aircraft and acquire the submarine by the centimetric radar. Some submarine commanders, who had survived such attacks, started to get wise to the situation and complained to their command that it was possible that the radio traffic was the culprit and not the Allied radar. The German command, however, under the belief in the Allies' awesome capabilities in electronics, decided with absolutely no basis for such a decision that the H2S had a useful range of fifty kilometers, which was in truth far beyond its capabilities. They told the submariners, in no uncertain terms, that they were wrong and would have to make do as best as they could in the meantime. Rommel too had a similar experience. On his retreat from El-Alamein, he was outmaneuvered too often by the British. The old Desert Fox started to smell a rat, but when he complained to Berlin, he was told that this was plainly impossible. Ironically, the "Rotterdam" find, while disclosing a lot of Allied radar technology and secrets, helped to protect the Enigma secret from a deeper investigation by the Germans.

To confuse the issue even more, the British leaked information that they were also using infrared (IR) to discover the submarines. R. V. Jones, who at the beginning of his career specialized in IR, was the originator of this deception. It was based on the fact that in natural light, a gray-painted submarine at sea appears gray to the eye and so does the sea. When viewed in the IR part of the spectrum, the submarine is still gray but the sea appears black. The Germans swallowed the bait, developed a sophisticated paint that looked gray in natural light and black in IR, and repainted all their submarines with this new paint. It was, of course, a total waste of time and effort, but one can understand the Germans' dilemma, even had they suspected that they were being led down the garden path. To paint was safer than to take a risk.

But still there was no end to the problems for the German submarines. In May 1943 the long-awaited Naxos receivers started getting to the submarine fleet, and nothing happened. Allied aircraft continued to locate and attack submarines at night, and the submarines usually were not warned until it was too late. Now the Germans became really worried. Was it possible that the Allies were using yet another unsuspected device to locate submarines? The truth was that the Naxos receivers were poorly designed and manufactured, with very low sensitivity. Furthermore, under pressure of the deteriorating situation, the whole installation of the receivers was shoddy. The proper way to do this was to mount the receiver's antenna on the conning tower and pass the cable through the pressure hull into the submarine. This was a critical piece of work, but apparently, in order to save installation time when the submarines were in port, the antenna was mounted with clips on the conning tower and a coaxial cable led from it, through the open hatch in the conning tower, to the receiver inside. In case of alert, everybody cleared the bridge and the last one down released the antenna clips, grabbed the antenna, and ran down pulling the cable after him. This kind of abuse soon caused bends and kinks in the cable that further degraded the incoming signal.

Somebody suggested that it was possible that the old Metox receivers, for the 200-MHz band, which were still in use, emitted some kind of telltale radiation. Could it be that Allied aircraft were homing in on this radiation? The scientists were asked about this and said no, but then the scientists had already been proved wrong once before when they said that usable centimetric radar was impossible. The Germans were still so influenced by that discovery that they were willing to believe anything about Allied capabilities in electronics. So a test was quickly organized and, lo and behold, the Metox *did* emit enough radiation for a good special receiver to home in on it. An order was immediately issued to shut down all

Metox receivers and an effort was made, totally unnecessary, to produce new and "safe" Metox receivers.

In an unusual turn of events, German records revealed that about a week after this order was issued, a British (RAF) prisoner of war admitted under interrogation that the British were using this kind of spurious radiation to locate submarines. Since this was a purely fictional story, the British made an effort after the war to locate the prisoner but to no avail, and even his name disappeared from German records. Two possibilities were advanced about the identity of this prisoner. One, that he was somebody with enough technical understanding who had figured out what the Germans were after and decided to give them what they were looking for. Another possibility is that the man was a "plant." According to this version, the Allies found out about the Germans' consternation over the Metox and decided to amplify it with a properly prepared volunteer who then let himself be taken prisoner. Considering the petty jealousies and in-house fighting that were permanent features of life in Germany, a third possibility can be advanced—that the prisoner, and the results of his interrogation, was invented by somebody in the German or French industry (which produced the Metox receivers) in order to discredit the Metox. Be that as it may, if there really was such a prisoner, he deserved a very big medal.

Now the Germans started rationalizing previous failures and connected them with the bogus Metox radiation homing. The largest number of attacks against submarines occurred in the Bay of Biscay, and not in the submarine patrol (and convoy attack) areas in the open ocean. The real reason was geographic. Because of its proximity and the fact it was somewhat of a bottleneck, patrolling activity by aircraft was denser in the Bay of Biscay than in the mid-Atlantic, and thus the chances of stumbling upon a submarine were higher. But the Germans reasoned that the relative scarcity of attacks in the mid-Atlantic occurred because after transiting the Bay of Biscay, the submarine crews disassembled the Metox antennas, which were also temporarily mounted, like those of the Naxos. The Germans also mistakenly concluded (again with no factual basis) that the Wellington bombers (which after having been withdrawn from bombing missions over land were used for maritime patrol) were too small to carry a centimetric radar, and their continued use was due to the ability to home on the Metox radiation.

Only in September 1943 did the Germans put their house in order and start operating efficient Naxos receivers. The three months that Watson-Watt had predicted would be necessary for the Germans to develop the required receivers stretched into eight months and cost the Germans dozens of lost submarines. The Germans also started suspecting Huff-Duff and tightened communications security,[8] but all was to no avail. The improved security and the improved Naxos receivers arrived too late to do the German submarines any good. In the summer of 1943 the "Atlantic Gap" was closed, and the convoys had constant air cover from aircraft, which took off from the United States, Iceland, Britain, and the new escort carriers that were now available in sufficient numbers. It is generally accepted that the "Battle of the Atlantic" was finally won in the summer of 1943, and from then on the submarine was transformed from a real threat to a diminishing nuisance.

Quite a large part of this victory can be attributed to the invention and the widespread introduction of the centimetric radars. A contributing factor to the Allied exclusiveness in this field was the longtime German belief that useful centimetric radar was physically impossible, and from here the assumption that what German science could not solve was in fact unsolvable. Ironically, when this mistake was discovered, the Germans went to the other extreme and convinced themselves that the Allies were capable of everything.

NIH (NOT INVENTED HERE) AND GUNNERY IN THE U.S. NAVY

Nobody likes upstarts and newcomers. They are least tolerated by an "establishment" that has it in its collective head that something or other is their private turf and nobody else should play on it, or alternately, because it truly believes it is the best organization to do the job properly. This of course can happen in any organization—

scientific, military, or administrative. Occasionally, in the field of technology this can happen in a situation in which an actual development project runs into difficulties and a useful suggestion is proposed by an "outsider." It is a rare occurrence when the outsider is made welcome and the idea is seriously discussed. In many cases the new idea is met with scorn and derision, and if the originator is persistent he runs into a wall of silence that quite often works. Eventually the person gets tired of the fight, and the whole thing is quietly and conveniently buried and forgotten. The organization has managed to salvage its reputation, and things can proceed again on even keel. Occasionally the outsider is stubborn enough or gets influential help, like Squadron Leader Leigh received with his light, and this is usually how progress is achieved. This of course can happen anywhere, anytime.

At the end of the nineteenth century, naval gunfire was still extremely inaccurate, and hits on target in a less than smooth sea were more a matter of chance than design. Because guns had limited elevation, they were more or less tied to the roll of the ship. The gun captain would look through the sights, and when he estimated that the roll was bringing the barrel to the right elevation, he fired. The rate of fire was thus limited to the roll rate, and accuracy depended a great deal on the gun captain's reflexes. A British naval captain, (later admiral) Percy Scott, came up with a possible solution called continuous-aim fire. It involved a change in the elevating gear of the guns, so that now the gun captain could continuously keep the sights on the target by rapidly elevating or depressing the barrel to compensate for the roll movement. Scott made the changes on his ship, the *Scylla*, and the results were quite extraordinary. Furthermore, he devised a training system using a rifle mounted in the gun and so could practice his crews almost without leaving harbor.[9] The British navy started adopting his ideas, but in the meantime he was assigned to duty in China. There he continued his experiments on his new ship, the *Terrible*. In China, Scott met an American, Lieutenant Commander William Sims. For years Sims had been concerned with the U.S. Navy's apparent inefficiencies and was quite outspoken about his opinions. The senior British officer and the junior American one struck up a friendship, and Scott explained to Sims his ideas about gun control. Sims was permitted to modify one of the guns on his own ship, the *Kentucky*, and in short order was achieving exceptional results. Sims then started writing reports to Washington about his experiments and his suggestions. The Navy Department simply ignored him. After another barrage of letters, Sims received the answer that the American equipment was as good as the British but the fault was in training, and this of course was the responsibility of the ship's officers (like Sims). Besides, he was told, his suggested solution would not work (it had been tried on land, which of course does not roll and this was important for the test), and anyway the problem had no solution. Sims apparently was not worried about his career (or he had powerful friends), so he wrote a detailed letter directly to the president, Theodore Roosevelt. The president ordered him brought back from China and made him inspector of target practice. To make a long story short, Sims ended as vice admiral and "the man who taught us to shoot."[10] Later, Sims was a staunch supporter of Billy Mitchell during the latter's court-martial and went as far as saying that the battleship was no longer a capital ship and that it would be replaced by the aircraft carrier.

On the whole, the Allies in World War II exhibited less of this too-human failing than did the Germans, possibly because of the shocking scare they received in the beginning of the war, which engendered a better cooperation between the various research and development bodies and the military. However, they were not completely devoid of such poor judgment. Recall the previously mentioned reaction of the Farnborough test center to Leigh's suggestion. With a little bit of charity one could almost accept the people of Farnborough's argument that they were on the verge of producing something better. However, upon closer examination it turned out that while their idea might have worked on land, they had no conception of the time and space problems of antisubmarine warfare.[11] The following case is obviously one of an organization trying with all its power to protect its reputation and possibly its employees' jobs.

PLASTIC ARMOR

One of the materials in short supply during both world wars was steel of all kinds. During the Second World War, the demand for the supply of steel of various types became critical due to the sheer magnitude of the conflict, the degree of mechanization of the participants, and the much higher attrition rates of equipment. Any material that could be used as a substitute for steel, in any application, was eagerly sought.[12]

Ships of war are constructed mostly of steel. Besides constituting the basic construction material, steel plating also serves as shielding for the protection of the crew in exposed locations. On the bridge and wheelhouse this steel must also be nonmagnetic so as not to affect the compasses, which at the time were north-pointing magnetic devices, and this type of steel was in limited supply. Various substitute materials were tried, including concrete, but it was found that when hit by bullets, concrete tends to chip and the ricocheting splinters are even worse than the incoming bullets.

In 1940, a navy officer named Edward Terrell, one of the members of the British Admiralty's DMWD (Department of Miscellaneous Weapons Development) came up with the idea of using a thick layer of bitumen, into which pebbles of crushed granite would be embedded.[13] This mixture would then be mounted on a thin layer of steel and serve as armor. The idea was that the incoming bullet or shrapnel would hit the granite pebbles and be deflected, perhaps more than once, without penetrating the backing. The bitumen would also stop the granite chips from ricocheting. Today this combination of a hard material in a matrix of softer ones would have been called "composite material," though at the time this term had not yet been invented. The man who invented the new substance constructed several such plates, took them to the department's firing range, and subjected them to strenuous tests. The results proved that the idea was perfectly sound. It should be pointed out that besides finding a substitute for the scarce nonmagnetic steel, there was the advantage of a much lower price. Bitumen and stone were almost free for the taking, and the only industrial resources needed for production were a mixer and a rolling machine. In England of 1940 that was a major consideration.

Terrell prepared a detailed report about the idea and the results of the preliminary tests and went to the Admiralty's Trade Division. They were the people charged with looking after the convoys and their ships. They were most enthusiastic and were willing to put the new armor plates into immediate production and mounting on ships. However, to do things in an orderly manner, in order to put new "construction" on navy ships, the Trade Division needed to get the approval of the Admiralty's Department of Naval Construction (DNC). This was considered a mere formality, and so the man who developed the concept, Terrell, and the chief of the DMWD, Commander Charles Goodeve, met with representatives from the DNC, presented the idea and the results of the tests, and suggested that if necessary the DNC should test it independently. Without any preliminaries, the DNC people said that there would be no test, and one of them explained to Goodeve that he had been in this profession for thirty years and he knew that the idea was no good, would not work, and was a total waste of time. On this note the meeting ended.

DMWD approached the navy's gunnery department, gave the gunners some plates of the new armor, and asked them to test them. The gunners were only too happy to oblige. They subjected the plates to exhaustive tests from all their weapons, and this being a bitumen-based product, also subjected it to flame. Their report stated flatly that the proposed material was better than any substitute yet proposed, and only the real thing—namely, steel plating—was technically better. Furthermore, they recommended that the material be adopted immediately. Goodeve now wrote a formal letter to the DNC, appended the gunnery department's letter, and asked them to approve the plates for use. The letter was returned with a curt response that the whole subject was a waste of time.

The whole affair was becoming a farce because in the meantime the gunnery department independently sent its own report of the tests to all departments in the navy, repeating their recommendation for immediate adoption of the proposed material. Goodeve, who was only interested in putting the new material into use, was willing to swallow his pride. He wrote another letter to the DNC, stating that he had destroyed his previous letter, which had been returned by the DNC. He now waited a few days and then re-sent a copy of that letter, with all the up-to-date accumulated test reports. In the meantime, the first sea lord came for a visit in the DMWD, and among other things the new armor plates were exhibited. Nevil Shute, the famous author, who at the time served in the DMWD, attached to the armor plate, by means of a pin, a note saying "PLASTIC ARMOUR—THE ONLY ARMOUR PLATE WHICH WILL TAKE A THUMB TACK!" The honored guests were duly impressed.

In the meantime, the DNC wrote to the gunnery department, complaining that the gunners had tested an armor plate that had not been properly developed or suggested by the DNC. After one of the senior people in DNC told Goodeve that they would remove their objections if the word "armour" were removed from the name of the new product, Goodeve better understood what bothered the DNC—although for a mature and experienced man like him, it was slow in coming. But now the Trade Division stepped in and insisted that the "armour" definition was important because of considerations of morale. At this stage the navy's higher-ups decided to ignore the DNC and go into production without their formal approval.

Every ship that came into port was equipped with the new armor. The design was passed to the United States and produced there for American ships. In the Mediterranean, it was produced in Egypt with hard porphyry instead of granite. The design was improved over the years, and this time around it was called "Plastic Protective Plating," and so even the DNC finally came around and removed its objection, this after a year's operational usage. It turned out that in some ways this material was even better than real steel armor plate. It better protected the crews from shrapnel, such as produced by mortar bombs. While ships were not usually subjected to bombardments by mortars, this did happen quite often in commando operations or during a landing on defended beaches. Such a bombardment was amply demonstrated in the raid on Dieppe, and later in all amphibious landings.[14]

Until the end of the war, the material served to armor some ten thousand ships, including most of the landing boats and ships that served in the various amphibious operations. It was later calculated that the new plastic armor saved over forty-four million dollars. The story had another happy ending. After the war, when the British Ministry of Defence distributed monetary awards to selected inventors who by their inventions contributed to the war effort, Terrell received some fifty thousand dollars, a princely sum at the time.

Because of the persistence of Terrell and Goodeve, the plastic armor episode ended in relative success. Nevertheless, the problem of NIH can occasionally evolve into considerably more dangerous and damaging situations. In the previous example, the price of stupidity could have been some unspecified number of dead and wounded and a certain waste of resources, but luckily the inventor was supported by his organization and presumably the behavior of the DNC people grated on everybody's nerves. The imprudence described in the following example did in fact cost a large number of casualties and did not end in total disaster, simply because by that time the Allies were superior to the Germans in practically every other field of weaponry.

THE SHERMAN TANK AND THE SEVENTEEN-POUND GUN

British Tank Development Difficulties

The British, who had "invented" the tank and introduced its use on the battlefield, failed quite miserably in tank development between the wars. There were several reasons for this state of affairs. Most of them were due to the attitude of the army as a whole and the lack of financing by the government, but there were also basic problems of management of technological projects. In 1919 the cabinet introduced the "ten-year rule," which assumed no British involvement in war for ten years.[15] It was thought that in an emergency there would be enough time to develop and introduce tanks. After all, during the Great War it was done in about a year.

Consequently, little money was allocated for tank development. Military writers such as J. F. C. Fuller and Basil H. Liddell Hart prodded the War Office, which produced some results, but very slowly. In the area of technical problems, there was a rapid turnover of managers of vehicle development and a poor use of manpower. An excessive number of prototypes, which were based on constantly changing requirements, were constructed. These requirements were issued by officers who not only knew very little about the problems of tank warfare, but were also incapable of, or not interested in, forward thinking. Because of poor financing, none of the basic prototypes could be produced in sufficient quantities to be formed into an armor unit, so that through a series of exercises their obvious flaws and useful features could be ascertained, formed into a coherent database, and incorporated into the following models. It should be remembered that at that time the level of practical experience in tank construction and tank warfare was very low, and the only way to generate such experience was by employing tanks in numbers, at least in exercises. The extent of the British army's poor approach to the question of tanks can be seen from the following quote taken from the British "Cavalry Training 1936" regulation: "The principle and system of Cavalry Training (mechanized) will be as laid down in Cavalry Training (horsed) with certain modifications laid down in this chapter. Mounted drill (in armoured cars) is based on the same principles as that of Cavalry. The principles of training in field operations given in Cavalry Training (horsed) are, in general, applicable to Armoured Car Regiments."[16]

The worst flaw in the process was probably the separation of the hull designers and the armament (gun) designers into two distinct groups. Even though the desirable tactical and operational modes of tank employment still eluded many at that time, the British did realize that a relatively big gun—and the bigger the better—would be an invariable feature of future tanks. Consequently, the gun's caliber and recoil system, the required quantity of ammunition, and the fire control system were the deciding factors of a tank's design. The hull and the power train had to be adapted to the main armament's characteristics. This necessitated close and constant cooperation between the hull designers and the designers of the weapon system and the turret, in effect melding them into one close-knit team, in which each side needed to be well aware of the problems and thinking of the other. These problems were compounded at that time by an additional factor. Today the function and the main armament of the tank are quite well accepted, and the designer of a conventional tank will most likely choose among the 105-, the 120-, or the 125-millimeter gun, the characteristics of which are well known. At the time under discussion, even gun design was very fluid and constantly evolving. The Germans, for example, discovered the potential of the eighty-eight-millimeter antiaircraft gun as an antitank gun during the fighting in the Western Desert. It had a heavy shell and a high muzzle velocity, ideal in the antiarmor role. The Germans used it then in this dual mode and later adapted it to be a tank's gun. The British also had an excellent antiaircraft gun, with a 3.7-inch (ninety-four-millimeter, in fact) barrel. Although this gun could serve as an excellent antitank gun and alleviate many of their problems in the Western Desert, the gun purists insisted that the role of this gun was antiaircraft and that it should be used against enemy armor only in an emergency.[17] In any case, this gun was probably too big for the period's tanks, so the tank designers had to create new guns for their tanks.

To be fair, it should be noted that the British did develop an excellent tank, the "Matilda," although it too was produced in insufficient quantities, and in any case they did not manage to repeat this success. This separation between hull designers and armaments designers led to a chain of difficulties that the British tank designers did not rid themselves of until they started receiving the first American Sherman tanks.

However, the good news was that the artificial separation within the development teams resulted in each team being composed of the better people in their respective fields. The gun teams were usually headed by artillery people, who understood the problems of gunnery (such as the relation among the shell's weight, muzzle velocity, and armor penetration) and were current in their state of the art. Spurred on by intelligence reports describing a new

German tank observed in a parade, they developed and put into production a gun that would fire a seventeen-pound (76.2mm) solid shell, with a muzzle velocity of more than twelve hundred meters per second (about four thousand feet per second). This performance was considered quite extraordinary at the time and was even better than that of the vaunted German eighty-eight millimeter, which achieved only 1,130 meters per second with a discarding sabot shell.[18]

The Problems of American Tank Guns

Although the Americans eventually developed an excellent tank, the M-4 "Sherman," and in such quantities that it could be also supplied to the British, the Canadians, and later to the French, from the start it was equipped with a very poor gun.[19] This was the short seventy-five-millimeter that, although it could fire an explosive shell (to be used against infantry), its low muzzle velocity, of about seven hundred meters per second, was too low to penetrate most German tanks. It should be explained here that for a given weight of shell, the ability to penetrate a given armor depends on the velocity squared. In other words, the British shell, comparable in weight to the American shell, did have a penetration capability nearly three times better than the American shell, and because of the higher velocity had a much flatter trajectory and thus was considerably more accurate. The United States Army Ordnance attaché in London regularly reported on the advancements made in German tanks, but his reports had to be forwarded to the States by the American military attaché in Berlin (before America's entry into the war). This man, who was a combat officer, did not consider the advancements important enough and did not endorse them.[20] Furthermore, the performance of American tank guns was tested in the United States against various American tanks and found satisfactory. So the armor experts in the States reached the curious conclusion that what was good enough to penetrate American tanks would be good enough to penetrate enemy tanks.[21] In the crucible of battle, it was proven that this was not so. When the facts became known, even General Eisenhower became angry and expressed his annoyance: "You mean our 76 won't knock these Panthers out? Why, I thought it was going to be the wonder gun of the war. . . . Why is it that I am always the last to hear about this stuff? Ordnance told me this 76 would take care of anything the Germans had. Now I find you can't knock out a damn thing with it."[22] The subject of Eisenhower's concerns is explained graphically in the Official History:

> The Shermans fought back desperately, stepping up to attempt to slug it out with their 75mm and 76mm guns, but the tanks that got close enough for their guns to be effective were quickly cut down by enemy fire. And when the American tankers did score direct hits on German tanks their shells ricocheted off the thick armor and went screaming into the air.[23]
>
> . . . The 76mm gun was better than the 75mm but did not have enough velocity to keep the tank out of range of the more powerful German tank guns, which were effective at 3,000 to 3,500 yards. . . . "The guns are ineffective, the crews know it, and it affects their morale," the tank commanders stated. They concluded the British had the right idea when they threw away the 75mm guns on their lend-lease Shermans and mounted their 17-pounders.[24]

A new gun, the ninety-millimeter, was finally developed in the United States, but in spite of its larger caliber, its performance was also less than that of the British seventeen-pounders. Furthermore, this gun could not be accommodated into the Sherman's turret, so it was eventually incorporated into a new tank, which started reaching the troops only in December 1944. While it is no consolation, it seems that the Germans had similar problems with their own ordnance people. After the fighting in France in 1940, it became evident that the German tank guns were barely adequate, and German tankers requested better guns. Hitler heard of the situation and ordered that the gun of the "Panzer III" be changed from a thirty-seven-millimeter to a long-barreled fifty-millimeter to achieve better muzzle velocity and penetration. But the ordnance specialists thought that they knew better. They argued that such a heavy gun would overload the tank in front and upset the tank's balance, and would also limit the tank's mobility in closed terrain.[25] So they put in a fifty-millimeter gun, but with a 2.10-meter barrel. This time Hitler was right, and when the first T-34 with a long barrel was captured, he liked to rub the ordnance professionals' noses in this fact. It took more than a year to straighten that matter out.

All this was bad enough, but armies occasionally find themselves unprepared for a war. What was truly unpardonable was the fact that a U.S. ally had a better gun, which had been offered to the Americans, but the U.S. had refused it.

The Seventeen-Pound Gun and NIH

A permanent British purchasing mission was stationed in the United States. When the British started receiving the Sherman tanks, they asked Chrysler, the producer of the Sherman, to substitute the British seventeen-pounder for the original gun. The British already had two years' worth of experience in true tank warfare, mostly in the Western Desert against the Italians and later against Rommel, and they knew quite well what worked and what did not. Without batting an eyelid, Chrysler agreed to make the necessary changes to adapt the gun to the tank. The British also offered the gun to the U.S. Army, suggesting that the marriage of the best gun in the world to a good tank would create a winning combination.[26] The American representatives politely thanked the British but refused. They also refused to copy the excellent German eighty-eight.[27] G. Macleod Ross writes: "This inaction by the U.S. Ordnance was in keeping with their predilection for quantity rather than quality, coupled with an arrogantly nationalistic spirit which refused anything which had not been invented in the U.S.A. (NIH—Not Invented Here)."[28] Ross adds that "as it turned out, it must have been intolerable to have your ally up-gun *your* tank and make it into a Tiger-mauler, but it was much more humiliating to win with quantity of inferior *materiel* and of blood, rather than with the quality which U.S. industry could have and would have furnished, if only its genius had been invoked."[29] This was NIH at its best, but the interesting question is, Who decided that this was intolerable—the British delegation's counterparts, or some higher authority, and if so how high? In any case, the British did not have inhibitions about British troops using American tanks (admittedly, at this stage of the game their only concern was about winning the war) and were quite happy with the combination they received.

There is another, slightly different version as to the failure of the Americans to adopt the British gun. Although without doubt the American ordnance people were well versed in the technical disciplines, they did not have any experience in the field and definitely no combat experience. On the other hand, the tank combat troops, including their officers, had absolutely no technical training and any ordnance specifications were totally incomprehensible to them.[30] This argument makes sense if we remember the aforementioned case of the military attaché in Berlin. If the enemy's armor thickness had no significance, why bother replacing a perfectly good, locally made gun with a foreign-made contraption?

All these problems had already started to come to light publicly during the fighting. In a *New York Times* news item from the front on December 4, 1944 (p. 4), men of the Second Armored Division stated that "the new 'Royal' or 'King' Tiger tank with its 'super 88' gun is the best tank in battle today. Our Shermans are all right in their class but they are outclassed." Finally, in another article in the *New York Times* (January 5, 1945, p. 4) Hanson Baldwin, a noted war correspondent, wrote about the shortcomings of American tank guns and the apologists in the rear who tried to explain their shortcomings. He in fact called for a congressional investigation of the whole affair.

It is really immaterial whether in this case it was a pure NIH sentiment, or a no less damaging technological "illiteracy" syndrome, which quite often and even today is a characteristic of otherwise quite capable fighting men.

THE SCANDAL OF THE AMERICAN TORPEDOES

The modern torpedo was invented in 1864 by Robert Whitehead, a British engineer living in Austria, who owned a successful business for the design and construction of marine steam engines. Whitehead hated war, and even more so after he witnessed some street fighting in Milan between Italians and Austrians in 1848. Later he met a retired Austrian naval officer who constructed a model of an explosives-carrying boat, powered by a clockwork mechanism and steered by two wires. Whitehead was fascinated by the idea and helped his friend, but after numerous failures they gave up. Whitehead then hit upon the idea that if the powered craft could run submerged and undetected, a terrible weapon could be developed that would help conserve peace. Being a marine engineer, his thinking was probably affected by his esteem of warships.

Whitehead started working on this project and eventually succeeded, while incorporating in his torpedoes a host of advanced ideas and engineering refinements. The motors in his torpedoes were driven by stored compressed air, direction was maintained by a gyroscope, and depth control by a device incorporating a hydrostatic chamber and pendulum, a feature that was retained in torpedoes until well after World War II. Austria, which at the time controlled part of the Adriatic shore and thus had a navy, was the first to purchase the torpedoes, the British were second in 1871, and by 1873 most other countries had followed suit. At the time, there were no real submarines in existence and the torpedoes were fired from surface ships. Whitehead torpedoes were first fired in anger in 1877, by a British ship, the *Shah*, against some Peruvian rebels. They missed, but their presence was enough to scare the rebels away. Because torpedoes did not require a big ship to be fired from, there emerged a new class of naval vessel—the torpedo boat—that participated in numerous naval actions and in effect threatened even major warships. This in turn led to the development of a "torpedo boat destroyer," which eventually became the "destroyer." Torpedoes were mated to submarines in 1900, for which they became the major weapon system. Torpedoes really came of age during the First World War, when they became an important weapon in the attempt to stop Allied shipping and starve England.

While not much could be done about protecting merchant shipping against damage from a torpedo, naval vessel designers had a little more leeway. All sorts of ideas were proposed to reduce the danger. These included netting lowered from booms (to snag the torpedo before it hit the ship, but which slowed the ship) or double hulls to prevent the explosion from breaching the inner hull (which were a good solution except that it could be adapted only to new construction). The double hull eventually evolved into tack-on additional armor plates (spaced away from the real hull) or "blisters" to protect only critical areas. Torpedo designers in turn were aware of all these proposals and came up with a counter of their own. A torpedo, after all, could be made to run at any depth. If a torpedo passed underneath the target and exploded directly beneath it, considerable damage would ensue. First, the bottom of the ship was less protected, and second, at the greater depth, the increased water pressure would amplify the damage from the torpedo's warhead, possibly leading to total breakup of the ship. The adoption of this idea might in turn lead to a smaller weight of the warhead, making a smaller and lighter torpedo possible. All this could be achieved by using a magnetic exploder, which would sense the steel mass of the ship (while disregarding local variations in the earth's magnetic field) and detonate at close proximity to it. Both England and Italy worked on magnetic exploders, although they eventually abandoned them as unreliable. The Germans experimented with magnetic mines during World War I and developed magnetic torpedoes and magnetic mines that were used in the Second World War, although they too had their share of problems with these torpedoes.

In the United States in 1922 a young naval lieutenant named Ralph W. Christie, who also held a master's degree in engineering from MIT, was put in charge of the development of the new magnetic exploder, or to use the navy's designation, "Magnetic-Influence Torpedo Pistol."[31] By the summer of 1924 a prototype was ready for sea trials, and a suitable target ship was needed to test both the working of the exploder and the effects of the deeper operating depth of the warhead. However, the naval BuOrd (Bureau of Ordnance) refused to allocate a ship for the experiment. Furthermore, BuOrd would not allow a torpedo carrying a live warhead to be fired from another vessel (only from the shore) because of the danger to the firing vessel. Remember, the whole purpose of the exercise was to prove that the torpedo worked, meaning that it was capable of sinking ships, and a live warhead was required for this. The torpedo-developing agency, the U.S. Naval Torpedo Station in Newport, Rhode Island, answered BuOrd by raising these arguments, and there ensued a lively exchange of letters that lasted for two years. Finally, in 1926 BuOrd relented and assigned an old submarine that was to be scrapped anyway. Newport was not happy with this choice, because the mass of a submarine is considerably smaller than that of a surface combatant, but that was better than nothing. The test took place in May 1926 and two torpedoes were fired from the shore. The first torpedo failed (it ran too deep), but the second ran true, passed under the target submarine, exploded, and the target sank. The test was considered such a success that further tests were canceled.

The new exploder was modified several times, and the Newport officials wanted to test it again. They also wanted to gather some data on the earth's magnetic field and data on the magnetic fields around "real" ships. The navy finally agreed to additional tests. In these tests the torpedoes, which carried dummy warheads, were rigged with an electric eye pointed upward, and when the torpedo passed under the target ship, the shadow of the target tripped the exploder mechanism without, of course, any explosives. There was a major fault in this kind of experiment. The performance of a magnetic exploder depends on its distance below the target ship, and such electric eye experiments really proved nothing in this respect. The electric eye would trip the exploder even if the torpedo ran much too deep. Nevertheless, everyone was enthusiastic about the exploder's performance, and Christie now requested a ship to carry out a full-blown test with a real warhead. However, the BuOrd refused to allocate a ship for the experiment unless the test team guaranteed to return the ship in good condition, and if the ship were sunk, the Newport team would have to raise it at their expense. This was patently a most unreasonable demand, but BuOrd was adamant on this point.[32] Christie finally gave up trying to subject the torpedoes to a real test, and approved the exploder for production in its present status; namely, without it ever having been subjected to a live test under realistic conditions. No additional tests were conducted to establish various other parameters such as effect of target's mass, proper storage and handling, and all the myriad tests that are performed on any complicated electromechanical explosive piece of ordnance. The manufactured exploders were stored under strict security, and a single copy of the manuscript of their maintenance manual was placed in Christie's safe without additional copies ever being made. The torpedo was designated as Mark XIV. To preserve security even further, submarine officers were not informed of the development of this weapon, nor how it was supposed to work, and were only told that the Mark XIVs were to be set to run deeper than the standard torpedoes.

In 1938 Christie was appointed head of BuOrd's torpedo branch, and one of his main assignments was to increase torpedo production, a problem that all navies suffered from. Deliveries of the torpedo started, but still without its exploder ever being subjected to a full live test and without the maintenance manual. The single copy of the manual was still buried deep in Christie's safe. It should be mentioned here that one of the considerations for a magnetic-influence exploder was the prospect of saving torpedoes in a real combat situation. A single torpedo exploding under a ship would probably sink her, while a torpedo with a contact exploder, hitting the side of the ship, might not always suffice and more than one would probably be required. In one final test in October 1941, just before shipments of torpedoes went to the fleet, it was discovered that the torpedoes were running a little deeper (by about four feet) than intended. This factor was not considered important enough to justify further testing, nor to inform the submariners.

When the war started, American submarines went to the Pacific and almost immediately sent back reports of an excessive number of misses and duds. Initial statistics have shown that more than half the torpedoes were fired to no effect. By now the submarine commanders knew what the new torpedo was supposed to do and how, and they complained that the torpedoes were running deeper than set and thus the magnetic exploder could not function. The submarine experts probably did not know it, but the electric eye tests finally came back to haunt them. Furthermore, they claimed that even when everything was just right, the magnetic exploder often failed to detonate. They suggested that until all these problems were solved, the magnetic exploder should be disconnected and the torpedoes used in the old mode, actuated by the contact exploder, which was retained. Admiral Thomas Withers, the commander of submarines in the Pacific, categorically forbade this change. After all, the name of the game was saving torpedoes. The result was that several of the submarine commanders disconnected the magnetic exploders on their own, swore the crews to a conspiracy of silence, and falsified their reports to make it appear as though they had used the magnetic exploders. When the number of "successful" engagements started rising, the Pacific submarine command waved this as proof that the magnetic exploder was not to be blamed after all, and that after the submarine crews learned to use it, all would be well. In the meantime, however, both the Royal Navy and the German navy discontinued the use of magnetic exploders because of poor reliability. This was known to the U.S. Navy but disregarded.

However, chief of BuOrd, Admiral W. Blandy, was becoming aware of the situation and as early as January 1942 warned Newport (the facility where the torpedo was developed) that something was wrong with the torpedoes; he even admitted that much to Admiral Withers. Withers, though, had already found out about the depth discrepancies in the October 1941 tests, and in April 1942 complained to Blandy about BuOrd's failings. In the meantime, Vice Admiral Charles Lockwood, commander of submarines in the southwest Pacific, who was constantly hearing of the torpedo problems from his people, decided to check matters himself. He purchased a fishing net, had it stretched vertically in the water, and had torpedoes with inert warheads fired at it. Lockwood found that on the average the torpedoes ran eleven feet deeper than set. These were the first reliable test results obtained after about eight hundred torpedoes had been fired in combat, and to Lockwood this corroborated to his satisfaction the submariners' complaints.

Lockwood reported his findings to BuOrd and in return received a letter claiming that his tests were "not scientific." Lockwood ran a second series of tests, which confirmed the previous findings—the torpedoes were running too deep. But this time Lockwood sent an additional copy of his report to the commander in chief of the U.S. fleet, Admiral Ernest J. King. This finally brought some response. In August 1942 Newport admitted that the torpedoes were running ten feet too deep and directed that ten feet should be deducted from the desired running depth. The situation was improving, but not quickly enough. In the tumult of the hunt for the correct depth setting, nobody paid attention to the magnetic exploder itself. These were the original designs from 1928, which had been created and produced using that era's technologies and had never undergone controlled testing. In the meantime, one of the more successful submarine commanders returned from a war patrol and reported that all magnetic exploders worked perfectly. BuOrd used this information to again claim that the exploders were perfect and that blame rested with the skippers (similar to Lieutenant Commander Sims and his guns). But Lockwood was fed up. With the approval of Admiral Chester Nimitz, commander of the Pacific Fleet, Lockwood had the boats under his command disengage the magnetic exploders and use only the contact exploders. Ralph Christie, the commander of U.S. submarines based in Australia, refused to do so. Since Christie (as noted) was the designer of the magnetic exploder, it can be assumed that he felt some sort of attachment to his creation.

The situation came to a head when a sub skipper came back and reported that he had made fifteen perfect shots at a Japanese tanker, using the contact exploder, and that all torpedoes hit the target but only two possibly detonated. It now appeared that the contact exploder had its difficulties too. Lockwood again took matters into his hands and organized a test. Three torpedoes were fired into a Hawaiian underwater cliff. One was a dud. A diver hooked a line to the torpedo resting on the bottom, and it was brought to the surface. Upon examination it was found that upon impact, and before it actually performed its function, the exploder mechanism was deformed to such a degree that the firing pin had been prevented from hitting the fulminate cap with sufficient force to detonate it.

Lockwood now subjected the contact exploders to a series of

controlled tests on dry land. It was found that a hit at or near ninety degrees to the target (considered ideal) resulted in a malfunctioning exploder. Only if the torpedo hit its target in a glancing hit did the exploder work properly. Admiral Blandy, at the BuOrd, still had difficulties accepting this, but an impromptu demonstration in the office, where a live exploder failed to detonate, finally convinced him too. It was later discovered that somewhere along the line, the direction of travel of the firing pin in the exploders had been changed, but again, without the new configuration ever being actually tested. In the meantime, submarine commanders were instructed to fire at their targets at oblique angles. Considering the difficulties of occasionally even getting into a firing position, it can be assumed that the sub skippers were not thrilled at this new directive.

In September 1943, "after twenty one months of war, the three major defects of the Mark XIV torpedo had at last been isolated. First came the deep running, then the magnetic influence feature, then the contact exploder itself. Each defect had been discovered and fixed in the field—always over the stubborn opposition of the Bureau of Ordnance."[33]

Robert Gannon lists several reasons for the depth-keeping difficulties.[34] Some were caused by poor management practices, but most were the result of poor design, which could have been avoided or rectified by either better theoretical understanding of the physical phenomena involved, or by very simple testing. The following are just three examples of such poor engineering (with some explanations):

1. For reasons of space, the depth control system, known as the Uhlan gear (a modified conventional pendulum control, which was originally developed by Whitehead himself), was placed at a slight angle to the torpedo's axis. Engineers believed that this placement would not matter. It did.
2. In response to continuing calls for "more punch," BuOrd added more and more explosive to the warhead, increasing it from 507 pounds to nearly 668. The center of gravity gradually shifted, altering running characteristics and adding

stresses beyond the capability of the Uhlan gear. This resulted in a deeper-running torpedo than had originally been planned.

3. Depth sensors were all tested in still water but never in moving water, so the calculations did not take into account the Bernoulli effect—that the side pressure on the torpedo's body drops as the velocity increases. To compensate for the side pressure drop, the torpedo went deeper.

Gannon relates another interesting observation. He quotes Joseph Henderson, director of the Applied Physics Laboratory of the University of Washington (which eventually worked on overcoming the exploder problems). Henderson claimed that the exploder was developed by the wrong type of people, which he terms "overgrown draftsmen." The "draftsmen" did not have the training of a scientist to test, test, and test until theory became fact.[35] What was true for the exploder was just as true for the rest of the design. It is true that the Bernoulli principle is somewhat esoteric knowledge, familiar only to those who are versed in fluid dynamics (aeronautical engineers, hydrodynamicists, chemical engineers, and certain specialized mechanical engineers). On the other hand, the concept of the center of gravity should have been obvious to anybody who ever flew a model airplane, sailed a boat, or designed a torpedo.

With regard to technical details, most of the difficulties did not stem from faulty or incorrect calculations. The problems started when people without the right background were asked to develop a very sophisticated technology, and they did not know how to deal with it and with the myriad scientific demands it posed. The fault was really with whoever assigned them this work in the first place. Then, when problems were reported from the field (or the sea, in this case), the people on the shore who could not deal with the facts simply refused to accept their existence, and this was the bigger failure.

The German submarine experts ran into almost the same problems with depth keeping and magnetic exploders. When German submarine commanders returned from patrols and complained about malfunctioning torpedoes, the German submarine command took them seriously. Admiral Doenitz appointed a committee of inquiry that initiated a thorough investigation of the torpedoes. Within weeks the magnetic exploders (which it was established could not be improved) were removed and an improved contact exploder was introduced (the original contact exploders had been accepted for service after only two trial firings).[36] The depth-control problem was eventually (in January 1942) traced to an air leak through a poorly designed gasket. The committee determined that four people were responsible, two civilians and two admirals. All four were tried for criminal negligence and punished accordingly. Lest it be thought that the Germans were just more efficient, at the trial it was disclosed that the problems had been discovered as early as 1936. Nothing had been done at the time to rectify them, the whole thing was swept under the rug, and it surfaced again only when the shooting started.[37]

In his exhaustive book about the U.S. submarine war against Japan, Clay Blair Jr. commented that "the torpedo scandal of the U.S. submarine force in World War II was one of the worst in the history of any kind of warfare."[38] Today we may agree that indeed it was a scandal, stemming from gross incompetence, poor management, and personal petty considerations, not in the least including misplaced feelings for one's own creation. However, a much later (1996) writer brings forth a curious argument.[39] In the beginning of the war, the U.S. Navy did not have many submarines in the Pacific. Holger Herwig claims that it is possible that if the torpedoes had worked properly from the start (and sank more Japanese ships), the Japanese would have been alerted to the submarine problem and could have developed countermeasures. This would have hampered the Americans when they finally fielded enough boats. As it happened, the Japanese were not prepared for the sudden American onslaught against their shipping later in the war. It can be assumed that Herwig does not imply it actually was better that the torpedoes were faulty, and thus did not warn the Japanese of the potential might of the U.S. Navy. But what Herwig does certainly say is that it can be ruinous to conduct offensive operations if one does not have an overwhelming force, and what he implies is that the U.S. Navy should have withheld operations until it had a much bigger force.

While it is certainly desirable to conduct war with the maximum of resources, Herwig's argument is mistaken on two counts. First, in the above case, if operations were intentionally held to a minimum, it is possible that the problems with the torpedoes would not have been discovered until a year or two later, with more faulty torpedoes in store. Second, in any new situation—and war is no exception—there is always a learning period. Doctrine is modified, weapons are adapted, and the better people are promoted. This time is priceless for those who know how to use it properly. No doubt the success of the American onslaught in the latter part of the war stemmed to a certain extent from the experience gained during the early years. Better torpedoes would have greatly improved that learning experience. As Blair commented on the same topic, "Intelligently employed, with a workable torpedo, submarines might have entirely prevented the Japanese invasion of the Philippines and the Netherlands East Indies. Skippers emboldened by swift and certain torpedo success, instead of puzzled and dismayed by obvious torpedo failures, might have inflicted crippling damage on the Japanese navy much earlier. The war in the Pacific might have been shortened by many, many months."[40]

"WINDOW" AND WHAT TO DO ABOUT IT

A good example of such an episode is "window," or as it is called today, "chaff." Window is a simple and effective antiradar invention that is useful even now against modern and quite sophisticated radar equipment. It consists of thin strips of aluminum foil cut to a length equal to about half the wavelength of the radar the strips are intended to counter. Dropped in bundles from the air, they disperse, causing an infinite number of echoes and in effect blocking the radar picture.

British scientists discovered this technique in April 1942. It was tested, found quite effective, and envisioned as a useful counter to the radars of the German air defenses, which plotted the RAF bomber raids. However, the employment of window was strictly forbidden on the grounds that it would be immediately copied by the Germans and used for the same purpose against Britain. Lord Cherwell, although usually in favor of any device that could aid the bombing effort against Germany, this time demurred. This controversy raged for sixteen months (and surprisingly enough, the secret managed to be kept) and was resolved only after the air staff calculated that had window been used for those sixteen months, more than 280 bombers and crews could have been saved. This number represented about one-quarter of the strength of the RAF Bomber Command at any time, and this argument finally brought Churchill to approve its use. In July 1943 window was employed in the first of a series of large-scale attacks against Hamburg, with resounding success. The British suffered less than 2 percent casualties on this raid, a ratio that was considered extremely low, compared with the occasional 5 to 7 percent losses. This first raid against Hamburg was followed by three more successful night raids by Bomber Command, and in between these, two daylight raids were mounted by B-17s of the Eighth Air Force.

Yet it was ironic that at about the same time that the British developed window, in the spring of 1942, German scientists also discovered the same effect and concluded correctly that it could be used to jam radars on a large scale. The Germans called the technique "Düppel." When Goering heard about it, he emphatically forbade any further development of the concept, reasoning that if the Allies discovered the effect, the results would be calamitous for Germany. Even discussing "Düppel" became a punishable offense. What's more, the Germans wasted nearly a year and a half without trying to find a counter to such possible discovery by the Allies on their own, which was exactly what happened. As previously noted, for a long time the Germans did not hold Allied technological capabilities in high esteem, until they were jolted from this complacency with the events of the centimetric radar, which were described earlier in this chapter.

The raids on Hamburg had another effect. These raids were conducted mostly with incendiary bombs. The old houses of this Hanseatic town—which contained much wood in their construction, as well as the huge amounts of combustibles found in any human dwelling—caught fire quite easily, and the widely dispersed fires soon coalesced into one huge conflagration. Large volumes of air, heated by this fire, rose upward, and cold air from the periphery rushed in, stoking the fires even further. This was the first occurrence of the so-called firestorm. Civil defense measures collapsed almost totally and the death toll came to some thirty to fifty thousand killed (sources differ on the exact number). It was the first time that civilian morale in Germany flagged to an extent that was noted even by the leadership.[41] In one interview, Willi Messerschmitt related that upon receiving the casualty reports from Hamburg, even Hitler, in a personal meeting with him, had confided that if raids of the magnitude of the Hamburg attack continued, "I will have to liquidate the war at once."[42] Luckily for the Germans, this fact did not become apparent to the Allies, and because Hamburg was considered sufficiently destroyed, the attentions of Bomber Command and of the Eighth Air Force were directed elsewhere. Window, however, continued to be used successfully by the Allies, while the German bombing effort against Britain was already waning and the use of Düppel would not have made a difference.

This example of window and the question of its timely introduction illuminates another important set of questions: How and when should a military technological advance be used or made known? Earlier it was noted that in the case of the Browning automatic rifles, Pershing was afraid to use them lest the Germans copy them (see chapter 3). There are many ways for this type of fear to be brought about. In peacetime a new military technology can be made known to the public as part of a public relations or political campaign; the latter is quite common in the democracies. It can be displayed in a technical exhibition or even advertised in an effort to sell it; in this way it is possible to recoup some of the development expenses or even lower the unit price in production. If the technology constitutes an extremely significant advance in weaponry (during peacetime, or at least not during an actual shooting war) should it be flaunted in order to intimidate the adversaries or impress the Allies? In doing so there is a danger that others will try to copy the said advancement, surpass it, or even develop a countermeasure to it. The recent developments in nuclear weaponry on the Indian subcontinent are a good example of such problems.

If the new development is of immediate use—say, during border skirmishing—should it be immediately used, thus possibly minimizing casualties, or should it be kept for a more critical opportunity, perhaps a full-fledged war? This is no hypothetical question at all. The "Blazer," a suite of reactive armor for tanks, was developed in Israel in the seventies. When its development was completed, a fierce debate raged in the Israeli Defense Force: should the "Blazer" be immediately mounted on tanks and thus protect them during border operations, or should its existence be kept a secret until a "real" war erupted, if ever? Apart from the moral and ethical questions, this is an interesting and practically insoluble problem in operations research, comparing possible numbers of casualties in border skirmishes and in full-scale future war.

In the case of window, both sides were intimidated from first use of this device by the specter of the adversary using it too. But both sides failed during all this time—more than a year in fact—to try to devise a counter to this weapon. Such a measure could have been achieved either by some tactical innovation, an organizational procedure, or even a technological solution. Both the Germans and the Allies had all these capabilities, and actually later in the war the Germans worked out some partial solutions to the window problem using all three approaches.[43]

NIGHT OPERATIONS AND THE FAILURE OF BOMBER COMMAND LEADERSHIP

While the British gave up unescorted, long-range daylight raids, the switch to night operations did not improve the odds very much and furthermore reduced the ability to find the targets, thus affecting the accuracy of the bombing. On the technical side, one of the most glaring design flaws of the British four-engine heavy bombers (such as the Stirling, Halifax, and even the advanced Lancaster) was the fact that they had no protection (in the form of weapons) on the lower side of the fuselage. In fact, they did not even have any observation in that direction. In comparison, the B-17 had nose (and later chin), top, bottom, and rear turrets plus two waist gunners, all except for those in the nose equipped with .50-caliber machine guns. To the traditional way of thinking in the British Bomber Command, the lack of bottom weapons was not a terrible disadvantage. Considering night conditions, the limits on night visibility, and the fact that fighters were equipped only with fixed, forward-firing cannons, an attacking night fighter usually approached from the rear, where he was exposed to the rear turret gunner and his four machine guns.[44] Furthermore, four crew members—the pilot, the bombardier, and the gunners in the top and tail turrets—were continually on the lookout for approaching fighters. Consequently, Bomber Command thought that adding another turret with its attendant weight and aerodynamic drag was a waste of effort and efficiency.

A German armorer in a night fighter unit saw the peculiar arrangement of guns on a bomber and hit upon a brilliant idea.[45] He took a pair of twenty-millimeter cannons and mounted them behind the cockpit of an Me-110. These guns were upward pointing and supported on a block of hardwood. He also fixed a periscopic sight for the pilot, who now approached his target from below, in the bomber's blind area. When in position, a few dozen yards below the bomber, flying steadily and at the same velocity, the pilot opened fire. His biggest worry was that pieces of the stricken bomber would hit his own airplane. The system proved so stable and efficient that the night fighter pilot could choose where to hit, preferably the wing with its fuel tanks, because a burst on the fuselage might explode the bombs carried there.

One of the more striking aspects of the story is the fact that an armorer was permitted to make a major modification to a combat airplane without a lengthy series of official tests to prove efficiency and safety. By the time of the raid on Peenemünde (August 17, 1943), two aircraft were thus equipped and each scored twice on that raid. Pretty soon the armorer was swamped with "orders" for his homemade innovation, which was officially named *Schrage Musik* (Oblique Music). The system was widely used and with great success.

Interestingly, the operations research group at Bomber Command headquarters suspected such a modification to German night fighters and alerted their superiors to it.[46] Also, some returning crews who managed to see the modified aircraft in action (presumably happening to their comrades) reported it during debriefing, but were not believed by the debriefing officers. Nothing was done about the situation, and Bomber Command completely ignored such a possibility, although it could have been easily checked out by simply modifying several aircraft and placing an observer to look down. This, however, was not the only case of poor judgment on the part of Bomber Command.

All bombers of all nations had escape hatches through which the crew could bail out in case the bomber was hit badly enough. The location and arrangement of these escape hatches differed from one aircraft to another depending on the particular design.

The operations research group at Bomber Command headquarters investigated the number of successful escapes from stricken bombers, by type of bomber. This calculation was based on prisoner-of-war lists provided by the Red Cross and on accounts of downed airmen who managed to avoid capture and returned to England. Obviously these statistics were approximate, but for purposes of comparison they were valid enough. A general trend was clearly indicated.[47] The number of successful escapes from B-17s was about 50 percent. From Stirlings and Halifaxes, it was about 17 percent and 25 percent, respectively. But from the newer, and more advanced Lancaster, only 11 percent of the crew escaped. Part of the reason for the discrepancy against the British bombers was of course due to the difficulty in getting out in darkness from a stricken aircraft—the B-17s flew on daylight raids. But this could not explain the difference between the older types and the Lancaster, since these all flew at night, sometimes in the same bomber stream. Finally one enterprising researcher took a measuring tape and investigated, and it was found that the width of the escape hatch on the Halifax and Stirling was twenty-four inches, while on the Lancaster it was reduced to twenty-two inches. Considering the difficulty of getting out through a narrow hatch while wearing a bulky flight suit and a parachute, this two-inch reduction in hatch size could explain the difference in escape ratios. The matter was brought to the attention of Bomber Command headquarters and was acknowledged in its official report.[48] It was also acknowledged that the Operations Research group had suggested improvements to the escape systems. What the official report does not say is that it took nearly two years, until toward the end of the war, before an enlarged hatch became standard.[49] While a design change, production, and retrofitting of an improved hatch may have taken a while and is partially excusable, doing nothing is not. Apparently, after being informed that there was a problem, for several months Bomber Command mulled over what to do, and in the meantime did nothing. Considering ongoing bomber losses, this two-inch difference in hatch width cost the lives of an unknown number of aircrew, possibly as high as several hundred.

Bomber Command's reaction to other reports from the field, about equipment difficulties that endangered aircraft and crew, was just as slow. (In fairness it should be pointed out that disregarding reports from the field was not only a British shortcoming. It has happened with the Germans and Americans too. Perhaps this is a malady of higher command, no matter what their national origin.) Consider the case of the "Monica."[50] When the night bombing raids against Germany intensified, so did the reaction of the German night fighters. Somebody came up with the idea of installing warning radar in the tails of the bombers to warn the pilot of the approach of the stalking fighter. While at first glance this seems like a good idea, as an extension of the ground radars, it had one major flaw with which ground radars can live but airborne radars cannot. With some generalization it can be said that the range at which a radar device can acquire a target is shorter than the range at which the energy broadcast from this device can be discovered. This is because the energy of the radar must go to the target and return, and still be strong enough to be received. The receiver in the night fighter receives the same energy after only half the trip of the broadcast. After an aircraft carrying this device was shot down, the Germans discovered this radar and installed suitable receivers in the night fighters. The warning radar actually became a beacon, yelling rather loudly, "Here I am." When the radar discovered the approaching fighter, it was already too late. While surviving crews complained about the possibility of the radar beckoning to the fighters, it took many weeks until somebody decided to seriously investigate, and of course these radars were immediately removed.

The same thing eventually happened with the ground-mapping H2S radar that served for navigation and target location. After the Germans captured it, they built receivers for the radar and installed them in their night fighters. Untold numbers of British aircrew died because the H2S helped the Germans to home in on the bombers.[51] The British, however, apparently refused to realize that this radar might act as a beacon and preferred to continue to use it in spite of the danger, possibly because it was the only device that enabled them to bomb at night, even when it was overcast. It is amazing how such lessons are not properly assimilated or even remembered. Ben Rich, in his book *Skunk Works,* relates a similar incident. The U-2, the famous spy plane, was equipped with a "black box" that was designed to jam Soviet air defense radars. When Gary Powers was shot down in May 1960 while flying over the Soviet Union, the possibility was raised that this "black box" served as a beacon, but the issue was never satisfactorily resolved. Several years later, a flight of four Taiwanese U-2s was intercepted over Mainland China and three were shot down. The fourth pilot marveled at his good luck, until at the debriefing it was ascertained that he had forgotten to turn on that jamming device.[52]

THE JAPANESE ATOMIC BOMB PROJECT

It is generally accepted that the first atomic bombs, dropped in August 1945 on the Japanese cities of Hiroshima and Nagasaki, came as a surprise to the leadership and people of Japan. It is also well known by now that the secret of the development of these weapons (the Manhattan Project) was not so much of a secret from the leaders of the Soviet Union, who were well informed of the progress of this development by their spies. It is now also well known that the Germans had their own atomic project, although for several reasons it was getting nowhere. To start with, there are now some claims that the German scientists were intentionally dragging their feet so as not to provide Hitler with such a terrible weapon. Also, it was claimed that the German scientists were afraid to proceed too fast because they were afraid of the consequences in case of failure. Apart from these claims there were some technical difficulties with the project. The German scientists themselves were not convinced that the whole concept of an atomic bomb, which would be small enough to be carried by a bomber aircraft, was technically feasible at all, and they might have been thinking more in terms of a power plant—for ships, for example. Their doubts stemmed from erroneous calculations that led them to believe that a greater quantity of uranium was needed than was actually the case.[53] They also mistakenly concluded that heavy water, rather than graphite, was the ideal moderator for an atomic pile.[54] These misjudgments probably would have been remedied in time, but meanwhile, not enough resources could be allocated in war-ravaged Germany for a truly meaningful research and development project in such a totally new field. In addition, the Germans were not unduly worried that anybody would beat them to it. They knew that the British had their own problems with resources, and they did not consider the Americans smart enough to succeed where they, the Germans, were barely making headway.[55] Misplaced self-confidence is a dangerous thing.

The Japanese case is no less instructive.[56] A Japanese general with technical training named Takeo Yasuda, who was the director of the Aviation Technology Research Institute of the Imperial Japanese Navy and who read much foreign technical literature, followed all the publications in 1938 and 1939 about atomic research. Seeing where all this could lead to, in 1940 he ordered one of his assistants, Lieutenant Colonel Tatsusaburo Suzuki, to prepare a report on the subject. Suzuki confined his report to the question of the availability of sufficient quantities of uranium, including occupied countries such as Burma (which was on the list of future Japanese conquests) and Korea, which had been under Japanese rule since the country's 1894 war with China. His conclusions, at least in this respect, were positive. Yasuda then turned to the experts and specialists, including Yoshio Nishina, who was a brilliant, western-trained physicist (he studied with Niels Bohr, the renowned Danish physicist who in 1922 received the Nobel Prize for his contributions to the quantum theory in atomic physics). Nishina and his students started some experimental work at the end of 1940, and in April 1941 the Imperial Army Air Force ordered Nishina to conduct research that would lead to the construction of a Japanese atomic bomb. This was very positive thinking on the part of the Japanese.

In the meantime, the Japanese navy also became interested. In the spring of 1942, an Imperial Navy committee proposed that research leading to an atomic power plant for ships be undertaken. Their thinking might have been influenced by the fact that the Japanese had never developed a decent logistics system for supplying ships at sea. Concurrently a secret committee investigated the feasibility of constructing an atomic weapon. The committee dealing with the project met numerous times, organized physics seminars, and tried to reach a decision. The decision hinged on answers to two questions: First, was the whole concept technologically feasible? Second, could Japan, in the middle of a major, difficult, and probably long war, spare the industrial and scientific resources to embark on such a project? The deliberations were probably influenced to a large extent by the fact that in the meantime the Battle of Midway was fought and lost, and Guadalcanal was invaded and for all practical purposes lost too. The committee's conclusion was that the atomic weapon project might take as long as ten years and would consume half of Japan's copper production and one-tenth of its electricity output. It was agreed that such expenditures would put an unacceptable strain on Japanese resources, and in any event, the present war would be concluded before such a weapon could be brought into use. The recommendation thus was to terminate the project and use the resources, particularly the people, elsewhere such as in the radar field. It is interesting to note that the committee also considered whether Germany, as an ally, and the United States, as the major enemy, were in a position to develop such a weapon. They were almost sure about the Americans because of the sudden drying up (in 1940) of American publications in the field. As we shall see in the next chapter, this "publications silence" was also noted by the Soviets. The committee's scientists concluded that in view of the present war's demands, neither Germany nor the United States had enough spare scientific and industrial capacity to begin such a project.

This conclusion was reached in 1943. While being historically interesting (and to students of military and economic intelligence affairs quite important), it really made no difference as far as the Japanese were concerned. Even if a completely opposite conclusion based on irrefutable evidence would have been reached concerning the Americans, the Japanese could not do much about it. After December 1941 it was simply too late. On one side were the feelings of the American public, who were out for blood and would accept no less than total and unconditional surrender. On the other side were the traditions and the opinions of the Japanese leadership, some of whom would have wanted to go on fighting, no matter what. They did plan to do so in fact even after the Nagasaki bomb on August 13, 1945. Four days after the second bomb and two days before Japan finally surrendered, Admiral Onishi Takijiro (who in October 1944 founded the first Kamikaze units and was now vice chief of the Imperial Japanese Navy) tried to dissuade the Japanese war council from discussing surrender. He offered to lead twenty million suicide fighters against the Americans.[57]

Nishina, who worked for the army (which in many respects was the navy's rival), continued working on a very small budget and even started some isotope separation experiments. But although his ideas were generally sound, he was getting nowhere. Finally, in April 1945 a stray bomb from a B-29 bomber burned down Nishina's laboratory, and that was the end of Japan's atomic project.

When the first atomic bomb detonated over Hiroshima, very few Japanese knew what hit them, or had even heard of atomic physics. One of the few who suspected an atomic bomb, immediately upon receipt of reports of the disaster caused by a single bomber at Hiroshima, was Lieutenant General Torashiro Kawabe, vice chief of the army general staff. He was probably involved in the original army efforts, hence his understanding. But until President Truman's speech a few hours after the first bomb was dropped, none of the Japanese decision makers even understood that for all practical purposes the war was over. Even then, some of the military leadership did not understand this almost certain idea.

6 Preconceived Ideas, Overconfidence, and Arrogance

The history of science properly does not concern itself with the things of science: the plants, the animals, the molecules, the atoms, the quanta, or even the law or the equation. The only object of study in the history of science is Homo Sapiens.

A. Hunter Dupree

All power tends to corrupt, absolute power corrupts absolutely.

Lord Acton

INTRODUCTION

Quite often, the personalities of the people engaged in work in the fields of science and technology have almost as much of an impact on the final results as the quality of the work itself. Since by definition almost any new development, important invention, or scientific breakthrough in some way threatens previous methods, states, or preconceived ideas, it is only natural that the proponents of the novel approach, whatever it is, will run into some difficulties. In extreme cases the reaction might be similar to the Church's reaction to Galileo, who talked himself into a cat's whisker from the stake, because the idea that the earth revolves around the sun was considered heresy. In our more enlightened times, the reaction to innovative suggestions could vary from denunciations in the professional literature to management's various and often imaginative replies as to why it cannot be done or why they cannot do it.[1] Here is where the personality of the originator comes into play. A forceful but diplomatic approach may help, although combinations of these two, together with a sharp mind, capable of original thinking, are few and far between.

With time, seniority, and experience, reputations are often established. If some real achievements are thrown in, such persons attain the position where their opinions, in their fields of expertise, and often on other general subjects, are sought after and acted upon. The utterances of such people of authority, in science or politics, can often set in motion (or stop) long-term trends, sometimes beneficial, sometimes less so. The statement made by Stanley Baldwin that "the bomber will always get through," quoted previously in chapter 3, is an example of such an episode. At the time, Baldwin was actually advocating global disarmament, but his speech brought about two different reactions. In Britain it helped to initiate the effort to bolster the defense against attacking bombers. In fact, during Baldwin's third term as prime minister (from June 1935 until Neville Chamberlain became prime minister in May 1937), he started the British rearmament effort. In the United States the devotees of the heavy bomber considered that statement a heaven-sent gift.

Not always do such statements have such far-reaching consequences, nor are they always made public. Nevertheless, it is interesting to note some opinions—at least as a reminder that unsupported statements or opinions are nothing more than that—even if made by a reputable scientist, and if taken too far, without proper foundation, can actually cause real damage. As we shall see, the above adage applies not only to politicians.

HENRY TIZARD AND THE BATTLE OF THE BEAMS

By the time the Second World War started, British officials had convinced themselves that they were the only ones who had radar, although everybody of significance was doing work in this field before the war.[2] The Germans and the Americans kept theirs secret, and since the British developed radar completely on their own, they decided that such developments could not have happened elsewhere. In fact the British got a very broad hint on German efforts in this field, but they chose to ignore it. In November 1939 a package sent from the British embassy in Oslo was delivered to R. V. Jones, head of scientific intelligence in the British Air Ministry. The package contained detailed information about various advanced weapon systems (and production capabilities) that the Germans were developing. It was written by a man who claimed to be an anti-Nazi German and included details about pilotless aircraft and guided bombs, about radars, proximity fuses, torpedoes, aircraft, and even mentioned Peenemünde as the site of some of the experimentation. This document became known as the "Oslo Report." Since there were some inaccuracies in the production numbers for aircraft, and since it seemed incredible that one person would have so much information, the British intelligence community dismissed the report as a plant. (Jones, however, believed it to be authentic, and as more information became available during the war, he kept checking it against the letter. After the war, Jones even located the writer, who was an electrical engineer at Siemens and because of his position was in constant touch with various German weapon designers.) So the British already had a hint about German radar technology and ignored it, but then they did something infinitely more foolish.

In December 1939 a British flotilla intercepted the *Graf Spee*, a German ocean-raiding "pocket battleship" (in essence a cruiser with bigger guns) off the shore of neutral Uruguay. After a short fight in which the Germans in fact gave more than they received, the *Graf Spee* entered the harbor of Montevideo. By the rules of war, the captain was required to leave in forty-eight hours or be interned (with ship and crew) for the duration of the war. The British made the German authorities believe that an even stronger force was waiting just beyond the horizon. This was achieved by the old ruse of the English flotilla signaling to nonexisting ships farther from shore and by phone calls made from the British consulate, which they knew would be tapped. To avoid internment or defeat in an unequal battle, the captain chose to scuttle his ship immediately outside the harbor. There was some mistake or confusion about the depth of the water, and when the ship sank, the water barely covered the decks, although the rest of the ship was badly burned. An enterprising British agent rowed to the ship and photographed her from all angles. He got excellent shots, including photos of the masts with what could only be radar antennas, which were for threat detection and ranging. In due time the pictures reached England, but it still took the Royal Navy several months to be convinced that the Germans actually had radar. It was a clear case of "My mind is made up. Don't confuse me with the facts." But the most important fact—that Germany had radar capability—refused to go away. The point, of course, being that compared to regular radio transmissions, the radar beam has to be focused into a very narrow angled beam. Such a narrow beam can also be used for other purposes; as a navigation aid, for example.

In June 1940 British intelligence started receiving information (supplied by Enigma intercepts, interrogation of downed bomber crews, and pieces of paper retrieved from crashed German bombers) that the Germans were planning to use some sort of very shortwave beams to navigate bombers at night and during inclement weather. The immediate reaction was denial of the possibility, since in the British opinion the Germans did not have that kind of capability in shortwaves, at about thirty megacycles or around a wavelength of ten meters. An additional objection was that since radio waves travel in straight lines, a beam originating in Germany would wind up too high over England (due to the curvature of the earth) to be of any use to a bomber. This latter objection existed in spite of the fact that there was a report by a noted British expert on radio waves propagation claiming that under certain conditions the beams could be detected at a lower altitude; in other words, the beams would "bend." Apparently one of the bigger doubters of German capabilities in this field was Sir Henry Tizard, who was instrumental in the development of the British radar effort. The question of such a navigation aid for German bombers was important enough to be finally brought before Churchill and the war cabinet to decide what to do about this threat.

The question of the beams came to a head in a meeting in Churchill's 10 Downing Street office on June 21, 1940. There were some ten participants, most of them high-ranking RAF officers and including, among others, Professor Frederick Lindemann (who was Churchill's scientific adviser), Hugh Dowding (Commander in Chief [CIC] Fighter Command), Charles Portal (at the time CIC Bomber Command), Robert Watson-Watt, and Henry Tizard as scientific adviser to the air staff. R. V. Jones, who had discovered and put together the information on the beams and was convinced of their existence, was also invited to this meeting. Jones later related that it was obvious that some of the people who voiced comments on this topic "had not fully grasped the situation."[3] Later he added that from the discussion it was obvious they did not know as much as he did on the matter, referring probably to the doubts about the very existence of such navigation and bombing aids. In due course, Churchill asked Jones to present his opinions on the subject. After hearing him out, Churchill accepted Jones's interpretation of the available data, and more important (from Churchill's point of view), his prediction that some means could be found to confound the German designs. On the strength of this information, Jones demanded the next day that a specially equipped aircraft be flown along certain routes to seek the transmitted beams, by listening on certain frequencies, doing in effect what the German zeppelin tried to do in the spring and summer of 1939 (see chapter 3). Luckily, Jones guessed correctly, and on the very same evening the beams were discovered and the doubters finally silenced. This eventually led to a massive effort in electronic warfare (the battle of the beams) and the later introduction of various electronic bombing aids by the Allies themselves.

One of the consequences of the June 21st meeting was that Tizard resigned from his position as scientific adviser to the air staff, in essence, the overall science authority of the Royal Air Force. Years later it was learned that inexplicably, Tizard had put his full professional weight into denying the beam theory and refusing to accept the evidence. It should be remembered that Tizard and his commission were the power behind the British radar effort, which they believed was unique (although this was before it was proven in the Battle of Britain). Maybe it was hard for him to believe that there were others, and mortal enemies at that, who had had the same foresight and were as good. When his opinion was finally rejected, he felt obliged to resign. Tizard later headed the famous "Tizard Mission" to the United States that established the close scientific cooperation between the two allies. One of the items the group brought with it to the States was the magnetron, the key to practical centimetric radar, which helped the United States surge forward in the microwave field. It might be instructive to speculate what would have happened if Tizard had had his way in that debate or if Jones had been cowed by his seniors, including the scientific ones. Tizard arranged for Jones's position in air intelligence, and Lindemann had been Jones's professor, although he kept his silence on this matter. It is interesting to note that both Lindemann and Tizard were test pilots during the First World War. Lindemann in fact developed the presently accepted method of recovering an airplane from a spin, which until then had been a sure killer. For a while, Lindemann was a member of Tizard's committee on air defense but consistently tried to push his ideas about detection of aircraft by their infrared emissions. The trouble was that the IR (infrared) technology was not yet ready for such a requirement. (In fact, even today IR is by several orders of magnitude behind radar in this respect.) Consequently, he did not get along with the other members of the committee and even tried to get Churchill to support him. Lord Swinton, secretary of state for air, ultimately accepted the other committee members' viewpoint, and Lindemann was finally eased out. When the committee's report was finalized, he again tried to challenge it, and it was only thanks to Air Marshal Hugh Dowding that the recommendations, to establish the Chain Home, went through. Lindemann, however, was a long-standing friend of Churchill and so became his adviser. It was pointed that had Lindemann had his way, the Chain Home radar would not have been built, or at least would have been badly delayed, and the Battle of Britain would have been lost.

Lindemann, who in the meantime was elevated to the peerage and named Lord Cherwell, and Jones were involved in another argument, this time on opposing sides.

LORD CHERWELL AND THE QUESTION OF THE V-2

At the end of 1942, British intelligence got its first inkling that the Germans were involved in the development of some rocket-propelled vehicle to serve as a bombardment weapon. Although the Oslo Report (received in November 1939 and finally accepted as genuine) mentioned both such a development and its location, in Peenemünde, this piece of information was kept on a back burner until more substantial information could be ascertained. Such further bits and pieces now started coming in, some of them even coaxed from the Germans by means of the Enigma decrypts, like the follow-up on the movements of a certain signals company (see below in the section about Enigma). But the evidence was confusing. There was almost no question that the Germans were developing some sort of an unmanned plane (which later was found to be the V-1), but the scientific establishment refused to acknowledge the existence of a long-range rocket weapon even when some photographic evidence started trickling in. This situation was aggravated by the fact that Lord Cherwell (Professor Lindemann) doubted the existence of such a rocket.

One of the problems in accepting the rocket as real was the question of compatibility of size and weight. Judging from reconnaissance photographs, the length of the rocket was estimated with a high degree of confidence to be about eleven meters (thirty-five to thirty-eight feet), with a diameter of about two meters (five to seven feet). Assuming that the rocket was based on solid fuel and considering the technology available then, the experts came up with a total weight of about eighty tons for the whole rocket. But this weight was patently impossible, because a rocket of the given size and weight could not contain enough fuel (the characteristics of which were approximately known) to lift such a weight off the ground and fly anywhere. Since this weight was reasonable for a solid-fuel rocket vehicle of that size, and because of this contradiction, the experts deemed the whole idea impossible.[4] In the meantime, it was established that there definitely was a pilotless aircraft project, and to be on the safe side, Peenemünde was bombed in

August 1943. There was some damage, and later it was found that the rocket project was actually somewhat delayed by the bombing, but in Britain the debate still raged on, with Lindemann leading the opposition. Finally it dawned on some people that instead of solid propellant it was possible that the rocket was propelled by a combination of liquid hydrocarbon and liquid oxidizer.[5] This should not have come as such a surprise to those concerned. In the United States, Robert Goddard had been experimenting with liquid rockets since 1925, and his work was well known all over the world. Some of the work of the German rocket experimentation group, of which Wernher von Braun was a founding member, was also well publicized (before the German army took over the program). Admittedly, since the German group openly talked about space flight, at the time they were considered a bunch of starry-eyed dreamers.

When liquid fuels came into the picture, some quick calculations were made and it became immediately evident that such fuels (in the V-2 these were liquid oxygen and alcohol) could supply enough energy to hurl the rocket some three hundred kilometers. This almost quelled the debate, although Lindemann still insisted that the whole concept of an accurate rocket to such a distance posed so many difficulties that he doubted the Germans would have attempted it at all. But the evidence was mounting up, especially after the D-Day landings, and the only unconfirmed question that remained was that of the exact weight. Eventually, the rocket turned out to be about twelve tons take-off weight with a one-ton warhead.

Again, had Lindemann's opinions been heeded, it is possible that the first raid against Peenemünde would not have taken place, or would have been made later. This would have enabled the Germans to start their use of V-2s earlier, almost concurrently with the V-1 attacks; and in the almost certain confusion and surprise at the new threat, this would have made the defense against the V-1 much less efficient. A somewhat earlier employment (even just before D-Day) of the V-weapons probably would not have won the war for Germany, not at that late stage. But in such a long, drawn-out conflict, victory is usually achieved on points, rarely by a knockout, and the points have a universal price—lost lives.

ENIGMA, PURPLE, AND OVERRELIANCE ON PRECONCEIVED NOTIONS

Communications are the nerve impulses in just about any human endeavor. In business, entertainment, and warfare they carry information and instructions between the various aspects involved in this activity. But while most human enterprises can manage most of their operations by means of the telephone line, the modern military cannot, or at least under most circumstances will be terribly hampered in doing so. Two simple examples are aircraft in flight and ships out of sight of land. Even a modern land army on the move will be limited in its operations if it has to depend only on telephones. Furthermore, since there are many demands on the use of telephone lines, particularly during operations, the end result will be delays that soon will be unacceptable. The invention (at the end of the nineteenth century) and wide introduction of the wireless communications system, the radio, was a great boon to the military too, and it was already widely introduced to the services during the First World War. Radios for a variety of uses progressed significantly between the wars, and the small German army of that period embraced radio use wholeheartedly, understanding that the wide use of wireless communications would help to quicken the army's operations. Reliable radio communications in fact were one of the cornerstones of the so-called blitzkrieg, the lightning war. Whether the blitzkrieg was real and defined as such by the German army, or a myth, as some historians now claim, the fact remains that radio was instrumental in facilitating the rapid communications between the various advancing units on the ground and with the supporting tactical air force. It is obvious that the German victories could not have been achieved that quickly without the extensive use of radio communications.[6]

The trouble with radios is that they broadcast in all directions, and anybody with a suitable set can listen.[7] (One day after the start of hostilities between Germany and Britain on August 5, 1914, the British cut the German transatlantic cable and thus forced them to use radio for their diplomatic traffic to America. The British listened to this traffic.) At a commercial radio station such broadcast power is useful, in fact desirable, but in military operations it may lead to a quick disaster, unless some form of hiding the real meaning of the message can be devised. Encryption and secret codes are of course very old concepts and were used from the dawn of history in letters and messages that were to be kept secret from unauthorized reading.[8] Generally speaking, the two main systems used to hide the meaning of a message are encryption and coding. Encryption is the substitution of the letters of a message by other symbols, letters, or numbers in a prearranged combination that can be designed with varying degrees of complexity. This enables messages to be written like any document, except that a straight substitution can be easily figured out, like in Edgar Allan Poe's story "The Gold Bug." In other words, it is not desirable to simply substitute an "A" for a "K," for example. The more effective way is that for the first time, the "K" will be an "A," but a second time it will be a "Z" and so forth, in some prescribed manner. The method in which this is achieved is actually the "key." Another possibility is to throw in a certain amount of additional dummy letters and enable the operators to know which are dummy and which are not. This in essence is encryption. The second system, coding, consists of the substitution of words and whole concepts by other words or numbers, creating in effect a dictionary. The two systems can be combined to further add to the difficulties of solving the problem; namely, reading the text without the benefit of having the "key" (the prearranged substitutions of the letters) or the "dictionary."

In consequence of the wide use of these methods, the art of the code breaker and cryptoanalyst flourished and to a large extent was institutionalized by the eighteenth century. Thus, when the military forces of the world became addicted to the use of the radio, they had to have some device or system to hide their communications from their adversaries, or for that matter from anybody other than the intended recipient. This of course applies just as well to diplomatic missions that are in constant communication with the home country, and doubly so during times of tension or crisis. Before the advent of long-range, efficient radio equipment, diplomatic missions and military personnel all over the world relied on commercial cable companies to transmit their correspondence (this fact was used by various governments to easily obtain copies of the secret traffic). One of the more bizarre cases in this field occurred on the morning (Washington D.C. time) of December 7, 1941. A note sent by the Japanese government to its ambassador in Washington, directing him to sever talks with the United States (and which was to be submitted to the secretary of state just prior to the attack on Pearl Harbor), was deciphered in Washington before it was submitted by the ambassador. But because of atmospheric disturbances, the critical warning about the possible start of hostilities could not be transmitted by the military radio network. It was instead sent to Pearl Harbor by RCA, a commercial company. Since no priority was assigned to it, it arrived in Honolulu at 7:33 A.M. (local time) and was sent by a messenger to army headquarters. In the meantime, the Japanese attack commenced, the messenger was delayed and finally delivered it at 11:45 A.M., and it had yet to be decoded.[9]

It was (and is) generally accepted, of course, that all codes and ciphers can eventually be broken, the only question being how long it will take to do so. So codebooks and keys (to ciphers) have to be regularly updated or entirely changed, distributed, and of course, kept from falling into the wrong hands. Code systems, however, have another problem. To be operationally flexible enough and thus useful, the number of such phrases in a codebook will be very large, and these volumes must be distributed regularly to all the units and headquarters that conceivably may have to use this code. For a far-flung army (like most armies in the Second World War, for example) this can become an insurmountable logistical task. This is exemplified in the logistic difficulties of the Japanese navy, which will be described later, and the events that led to the Battle of Midway—a crucial engagement that changed the course of the war in the Pacific.

Encryption Machines

In war, however, equipment is lost and prisoners are taken. Any encryption system must be capable of being used safely, or with very slight standard modifications, even after such an occurrence. What it boils down to is that there is a "system" that one can assume the enemy knows about, and there are "keys" that make the use of the system safe from the enemy. In a way, this is like a lock and a key. Unless you have the right key with the right combination of teeth, knowing the principle on which the internal structure of the lock is based is of limited use. However, as already stated, to have reasonable overall security in military encryption, it is desirable to change the keys or codes after a certain elapsed time, which may be hours or days or weeks. During World War I all these problems came to light, and it became obvious that one could have either easy wireless operation but without good security, or good security but with exceedingly cumbersome and slow procedures. Suffice it to say that the general worldwide trend in this field tended toward encryption (as different from coding), and several inventors pursued the idea of a machine that would automatically encipher messages by letter substitution.

By the year 1920 such machines were invented on both sides of the Atlantic, essentially based on similar mechanical principles. The most notable ones were by Edward Hugh Hebern in the United States, by Arvid Gerhard Damm in Sweden, and by Hugh Alexander Koch in the Netherlands. The Damm product was later acquired and improved by Boris Hagelin and sold extensively in Europe. The American product was technically good, and some of these machines were bought and used by the American military. Commercially, however, the venture turned out to be a flop, and eventually the firm went out of business. The United States finally settled on the Hagelin machine, and Hagelin became a millionaire from his sales to the United States government. The Dutch machine suffered essentially the same fate as the Hebern product, and in 1927 all patents and drawings were sold to Arthur Scherbius, a German engineer who thought he could perfect the encryption machine. He improved on the Dutch ideas and called the machine

Enigma, but commercially this venture also failed. The company changed hands a couple of times until in the early thirties the German army became interested and acquired several such machines. With Hitler's rise to power the new Wehrmacht (German armed forces) was expanded enormously, and from then on the Enigma was supplied in growing numbers to the German military.

The Enigma

The Poles and the Germans were longtime enemies who had fought several times over the centuries, and the intelligence services of both countries kept a wary eye on each other. In 1932 the Polish secret service became aware of Germany's reliance on the Enigma and sought to learn its secrets. They first obtained a commercial version, and a group of mathematicians tried to devise some statistical methods to decipher the output of the machine. This involved building a copy of the supposed military Enigma version and the development of some statistical theories. The group made considerable progress, but it became obvious that a full solution was still far away.

The basic operating principles of the Enigma were basically comparable to all encryption machines of that period. The Enigma was essentially similar to a typewriter except that the keys were arranged somewhat differently and there were only twenty-six keys for the letters in the Latin alphabet. Numbers and other symbols were spelled out and only capital letters were used. The machine contained several wheels (or rotors) with electrical connections, and three of these were inserted (in a random order) into the machine parallel to each other. The starting positions of the wheels relative to the machine and to each other could be changed. When operating the machine, pressing an "A" on the keyboard sent a current in a complicated and ever-changing path and finally lit one of twenty-six lamps. The letter corresponding to that lamp was the one to be sent by the communications device, be it Morse or voice. There was one more twist—the first wheel moved one space every time a letter key was depressed, and when it finished one revolution, the second wheel moved one space and the first wheel started over again, much like an odometer in a car. The combination of the random positions and initial setting of the wheels, their wiring, and the internal wiring of the machine gave more than 200 trillion possible, mostly random, substitutions for each letter to be encrypted.[10] No doubt the Germans were quite proud of themselves and felt it would be impossible to break the encryption carried out by these machines.

The keys for the setting of the wheels were changed every twenty-four hours, and a package of keys for one month was distributed each time to regimental headquarters (or the equivalent) and higher and to individual submarines. The daily key for the operator to set consisted of the following items: 1) which three of the five different wheels were to be used; 2) the order of the wheels to be installed in the machine; 3) the setting of the letters on the wheels, relative to the internal wiring of the wheels; 4) details about changes of the internal wiring. When broadcasting, the operator was free to choose the initial position of the wheels, and this information, plus the call signs of the sender and receiver and the type of traffic (operational, administrative, or training), was broadcast in the clear, followed by the enciphered message. After hearing his call sign and the other data broadcast in the clear, the receiver adjusted his machine according to that particular traffic key for the day and other pertinent key data and proceeded to punch the incoming message on his Enigma. If and when everything went right, the machine blinked back at him the decoded message. The main visible problem with this machine was that in order to be efficient three people were required to operate it. There were also a number of hidden problems that the Germans had not fathomed.

The Breaking of the Enigma System

In the beginning of 1932, a German clerk sold a file of decoded messages to a French intelligence officer, but the Frenchman's superiors were not interested. On his own initiative the Frenchman offered the messages to the Polish secret service, with whom he had good working relations. The Poles appreciated the gift, and the above-mentioned mathematicians were then able to finish their decoding machine and check their theories. In 1933 the Poles began routinely decoding German military communications traffic. In 1938 the Germans changed the wheels and added two more, and the Poles had to start practically from scratch, although most of the basic procedures remained valid. At the beginning of 1939, the Germans started distributing the new Enigmas to their units, and according to most published accounts (sources vary on this point) the Poles staged a raid and snatched one from a German military truck. They then substituted an old commercial Enigma and burned the truck. This was before the war. The Germans, who found some debris that ostensibly belonged to the Enigma, apparently fell for the ruse, and the Polish secret service started making progress again. In August 1939, feeling that war was near, the Poles called in their French and British counterparts and presented them with copies of the machines, the decoded messages, and the theoretical work performed to date.[11] The British and the French were dumbfounded but took the machines and hurried away, and the French machine eventually also reached London.

To make a long story short, the British finished the work, contributing a great deal of original thinking to the effort, and managed to get to the point where Enigma intercepts were being read within days, sometimes within hours, of being plucked from the ether. The Germans apparently made several major, and a host of minor, mistakes in their use of the machine. To start with, the number of 200 trillion possible permutations, although basically correct, was in fact misleading. Even one-hundredth of this number would have sufficed at that time, before the development of computers, to ensure safe communications if all other basic rules of communications security were observed. So the British cryptoanalysts had to revert to other means than just trying to run through all the combinations of letters in order to read the messages. As noted, they did build copies of the machine. The Germans probably guessed that much since they figured some of these machines would get lost in combat. The British also set in place the organization and procedures necessary to do the work in an orderly manner. This was basically an elaborate administrative setup to listen to the various German broadcasting stations, take down the intercepts, and note all the important details, such as call signs, frequencies, and so forth. Also, because the Germans trusted the Enigma implicitly, they used it for everything, including administrative work, political speeches, awards of medals, and all sorts of other frivolous traffic such as reporting from a submarine about a toothache. David Kahn, in his detail-rich book, comments "that the U-boat command became the most gabby military organization in all the history of war."[12] This huge amount of traffic enabled the British cryptoanalysts to get a "feel" for the various German operators. Also, the Germans tended to be very formal in addressing their superior officers and in signing their own dispatches. This, coupled with the call signs, the very orderly format of the messages, and the large amount of traffic, betrayed recurring patterns with little variations, which helped in the deciphering work. Furthermore, in time the German operators tended to get careless and lazy, changing wheel starting positions very little or not at all. When all these facts were added up, and with much experience because of the large volume of traffic, it turned out that occasionally the number of machine settings and permutations to be tested shrunk from trillions to dozens and less.

A broken cipher is understandable and should be expected as part of the communications war.[13] Based on this assumption, various methods and procedures should be devised and enforced to reduce the risk of this happening, or to mitigate the damage if it does happen. In view of this basic rule, the Germans' faith in their ciphers was at the very least puzzling. On the other side, the Allies were extremely careful in handling the decrypts from the German traffic. They instituted a rule that no such information was to be used unless it could be independently corroborated by another source. Obviously, knowing where to look facilitated finding such secondary sources. Also, the fact that a large amount of the information had no direct operational effect and was used for other purposes may be not so dramatic but still important. R. V. Jones relates how he wanted to know if the Germans were tracking the V-1 tests in the Baltic Sea by radar. He asked that an Enigma watch be put on the movement of certain of the better German signal companies.

When one of them was moved to the Baltic area, his suspicions were confirmed.

Lest it be thought that all was clear sailing for the Allies, it should be noted that on some occasions the Germans did introduce changes into their machines, and this of course was apart from the regular key changes. They changed wires, wheels, and in the middle of the war the German navy added a fourth wheel to the machine. However, by then the Allied, mostly British, cryptoanalysts had so much experience that they quickly weathered these minor setbacks. They were occasionally supported in this task by the capture of deciphering keys that the Germans did not succeed in destroying.

Japanese Overconfidence

The United States Navy started reading parts of the Japanese navy communications traffic prior to 1941, but their success in this effort was too fragmentary to avoid the Pearl Harbor debacle, although as pointed out above, the diplomatic codes were read almost routinely.[14] From the time of Pearl Harbor, the United States intensified its effort to break the Japanese naval messages and reached the stage where it could read most of them. The Japanese had no inkling of that activity, and in fact were very sure of the security of their traffic. They were, however, getting worried about the length of time that the present code was in use and decided to change it on April 1, 1942. They apparently did not appreciate the magnitude of the problem of distributing the new codebooks to all the units in their new far-flung conquests, including to ships at sea. For the Japanese these were the very successful first months of the war, and they found themselves suddenly in control of vast territory. To this challenge were added various administrative difficulties, and the Japanese decided to postpone the starting date for the new code to May 1. This plan did not pan out either, and in the meantime the entire Japanese navy's business was conducted by the old code and read just as quickly by the United States's cryptoanalysts. Thus the attack against Port Moresby in New Guinea was divined in time. This precipitated the battle of the Coral Sea, which, although it may be counted as a Japanese tactical victory, was the first real setback they suffered in the war. The Japanese masterstroke against Midway Island was also grasped early enough, and a proper trap was set for the Japanese task force. The Japanese finally managed to change their codes on June 1 (and it was several weeks before the American code breakers made any headway on the new code), but by then it was too late for the Japanese fleet. The Battle of Midway was fought on June 4, with the Japanese navy badly mauled, losing four carriers (to one American), and it signaled the beginning of the end for Japan. Had the Japanese changed their codes in April, or even May, all the critical intelligence that led to the Midway victory would have been lost to the United States Navy, and quite probably the battle as well.

The Ambushing of Yamamoto

The Japanese failed miserably in keeping their secrets from the American cryptoanalysts. While this kind of failure was almost regular and led among others to the lack of success in the Coral Sea and to the crushing defeat in the Battle of Midway, there was one other noteworthy Japanese failure. Although it is doubtful that it made a big difference at this stage of the war, nevertheless, it had military significance and without doubt a bad effect on Japanese morale.

On April 13, 1943, the Americans deciphered a Japanese message detailing the itinerary of Admiral Isoroku Yamamoto (the architect of the Pearl Harbor attack), who was planning an inspection tour of Japanese positions in the Solomon Islands in the South Pacific. The trip was to take place in five days. Although there was some doubt about the desirability of shooting down Yamamoto's plane, because doing so could disclose the fact that the Americans were reading Japanese messages, Admiral Chester Nimitz decided to take the chance. He was willing to take this risk for an additional reason. Previous experience showed the Japanese occasionally changed their ciphers or tightened security a little, but by now the Americans were sure of their ability to crack anything the Japanese could put up. So even if the Japanese changed their codes, it would be only a matter of time until the new codes would be broken, and presumably the damage would be minimal. The raid took place as planned on April 18. Eighteen P-38 twin-engine fighter planes took off from Guadalcanal, made the intercept over Bougainville, some 435 miles away, and Yamamoto's plane was shot down and burst into flames over the jungle. Despite an escort of Zeros, only one P-38 failed to return. Considering the effect of the removal of a very able enemy leader (who it was assumed could be replaced only by a lesser individual), the American naval command considered this coup equivalent to a major naval victory.

Overconfidence or Arrogance?

Just as for the Germans, the question was often raised, Why did the Japanese show such a lack of basic common sense in their communications security? One explanation rests on the recurrent Japanese failures to penetrate American ciphers. While occasionally they managed to read some messages in the simpler codes, they got absolutely nowhere in the more elaborate, and important, military and diplomatic ciphers. It is likely that their failures in this field persuaded them that successful cryptoanalysis, on a regular basis, was manifestly impossible. It seems that they projected this failure onto the Americans and concluded that their own ciphers were also safe.[15] Kahn relates that in April 1941 the Japanese foreign office was informed by the Germans that Japan's codes were being regularly read by the Americans.[16] The Japanese foreign office became worried and questioned the Japanese ambassador to Washington on the matter. The ambassador replied that "the most stringent precautions are taken by all custodians of codes and ciphers," and that was essentially the end of this episode. Apparently, the Japanese could not imagine that secret traffic could be read without stealing or copying the original codebooks, a somewhat naïve belief at best.

The Germans did not have even this excuse. They broke Allied ciphers quite often and used these decrypts, the revealed messages, to good effect. The Germans did not consider a code or an encryption system as inviolate. Nevertheless, they did consider their own as such.

This problem is puzzling for another reason. The Germans read the Allied transmissions to their convoys regularly enough. Some of this traffic included directives on the best routes for convoys to avoid submarine wolf pack concentrations. The Germans should have understood this rerouting business and should have seen that the Allies seemed to know too much about the wolf packs' locations. Allied air activity was a possible excuse for the Germans, but it was not good enough and they should have asked themselves how the Allies possessed this detailed information. However, the Germans never lost faith in the Enigma, or perhaps it was arrogance, an attitude that "No Limey can decipher what a German wrote."

THE PREDICTIONS OF IRAQI MISSILE ATTACKS

The next case described turned out to be a national joke in Israel, but under slightly other circumstances could have had dire consequences. In the tense days between Saddam Hussein's invasion and brutal conquest of Kuwait (in August 1990) and the start of Operation Desert Storm (on January 17, 1991), the political situation in Israel was in turmoil. Although in no way involved in the present crisis, Israel was a longtime ally of the United States. On the other hand, Israel had had its differences with its Arab neighbors and other Arab countries of the region, including Iraq, which is the only country that refused to sign the cease-fire accords after the 1948 war. What's more, Israel adheres to a long-standing policy that if attacked it will retaliate by suitable means. During the Gulf crisis, the United States was trying to meld a coalition, led by itself, of western countries, chiefly the U.K. and France (with material support from many others), and including regional powers such as Saudi Arabia, Egypt, and even Syria. The main reason for this complicated political arrangement was to emphasize that this was not an attack by the west against an Arab country, which could have had extremely disagreeable connotations, but a multinational effort, including by Arab countries, against an aggressor, even though he happened to be an Arab.

Into this cauldron of Middle East politics, rife with crosscurrents, stepped Saddam Hussein, who announced that in case of a war against Iraq he planned to use his missiles to "burn off half of Israel." Contrary to the case of the V-2 in World War II, where only the British leadership knew of the impending threat, in the nineties no government could enjoy such convenient secrecy, especially with all the freedom the mass media representatives enjoyed. In any case, Saddam Hussein wanted this threat to be known, so censorship would not have helped. Naturally, the Israeli media reacted profusely, although for a change the politicians were a little more guarded. The general public, torn between calls for a preemptive attack (such as the very successful 1980 Israeli attack against the Iraqi nuclear reactor) and counterarguments that this was exactly what Saddam wanted, was a little confused. The Israeli government played it safe and stepped up the effort to equip everybody with gas masks and protective equipment for children, and disseminated information about sealed rooms and other civil defense measures.

In the beginning of January 1991, about a week before the start of Operation Desert Storm, an "expert" appeared on national TV, and later that month expounded his opinions in an Israeli newspaper. On the face of it his credentials were impeccable. This was a reserve brigadier general, formerly in a senior position in military intelligence. He held a Ph.D. in international relations from a renowned European university and at different times sat on various strategic-planning committees in the defense and industrial communities. His message to the people was simple: "I promise you, with a 100 percent, not 99 percent, confidence that there will be no Iraqi attack against Israel. There is no danger from that quarter. In order to strike us, they have to initiate a war. To Saddam Hussein it is clear that his only chance to survive is to avoid a war. In a war he will be finished." Apparently, Saddam was too busy to hear that broadcast, or he failed to be impressed by it. On January 18, 1991, one day after the start of the American air offensive against Iraq, he fired six extended-range Scuds against Tel Aviv and several missiles against Saudi Arabia, and afterward kept on going even while his army was being smashed to smithereens. All told, some thirty-nine missiles were fired against Israel between that date and February 25. A similar number was fired against Saudi Arabia. Israel did not retaliate after all, so as not to endanger the cohesion of the coalition, but it could afford not to retaliate because politically the government of Israel was simply lucky. Considering the number of missiles fired into densely populated metropolitan areas, the casualties were extremely light. Two people were killed directly by a missile attack, five were dead from heart attacks attributed to the missile attacks, and seven people died because of the misuse of emergency equipment. In addition, some two hundred people were injured. So there was no public demand "to do something." Ironically, because of much lighter road traffic during the emergency, the death toll due to traffic accidents fell off sharply.

The question now is what induced the man to make such a statement. It is possible that he was put up to it by somebody in authority who wanted to counter the effect of all the preparations for chemical warfare, but nothing of the sort ever surfaced. So the only other explanation is the "Microphone-in-Hand Syndrome," where someone given a microphone feels he must say something, even if he has nothing to contribute. Certainly this expert is in the good company of other influential people who made statements without factual evidence, including among others Stanley Baldwin; Henry Tizard; Lord Cherwell; Charles Portal; and Vannevar Bush, who because of his outstanding position and influence deserves special attention.

TECHNOLOGICAL PREDICTIONS ABOUT FUTURE CONFLICTS

Introduction

Dr. Vannevar Bush (1890–1974) was a man of many and varied accomplishments. Trained as an electrical engineer and mathematician, during the First World War he was involved in submarine detection and ballistics research. He also headed the team that developed the first electronic analog computer. Later in his career, he was the dean of engineering and vice president of MIT, the president of the Carnegie Institution, and eventually in 1939 he was appointed as chairman of NACA, the National Advisory Committee for Aeronautics, NASA's predecessor.

Bush was popular in Washington, and he used his connections to lobby for the establishment of a national body that in the event of war coming to America would organize the country's scientific and technological effort to develop advanced weaponry. The fall of France doubtless convinced many in the United States of the probability of war and most likely was the trigger that brought President Roosevelt to establish the National Defense Research Committee (NDRC). Bush was appointed as the chairman of the NDRC. A year later, in June 1941, President Roosevelt established the Office of Scientific Research and Development (OSRD). Bush became head of the OSRD, while the NDRC, headed by James Conant, the president of Harvard and Bush's deputy in the OSRD, became the executive arm of the OSRD.

Eventually the OSRD became the world's most powerful scientific body dealing with implements of war. It controlled the work of nearly thirty thousand scientists and engineers in 775 institutions and financed some twenty-five hundred research projects. These included weapons, amphibious trucks, proximity fuses, acoustic torpedoes, the whole plethora of radars, aircraft rockets, military medicine, and even the supervision of the transformation of penicillin from a laboratory curiosity into a cheap product available for all. When the "Uranium Project" was started, it initially was part of the NDRC, but in mid-1942 it was separated from the NDRC and became the "Manhattan Project," directly under OSRD control. That alone cost more than two billion dollars (in 1945 dollars). The total expenditure of the OSRD is unclear, but some authorities place it as high as ten billion dollars (in 1945 dollars). Bush commented later that the U.S. Congress was giving OSRD appropriations in lump sums and trusted it to decide on what projects to spend the money, and there is no question that the money was well spent. Never before were so many scientific and financial resources concentrated in the hands of a single scientific administrative body, and under the able leadership of Bush they contributed enormously to the organization and tapping of America's scientific and technical potential. What's more, many of the war's developments contributed to the United States's long-standing world leadership in technology. An important industrialist said later of Bush, "Of the men whose death in the summer of 1940 would have been the greatest calamity for America, the President is first and Dr. Bush would be second or third."[17]

In September 1949 Bush authored a book in which he detailed his predictions about future warfare, in view of the recent advances in weaponry.[18] The book deals with a possible armed conflict with the Soviet Union and contains segments of text that enumerate the virtues and benefits of democracy. While nobody in his right mind would argue with that, Bush comes to the startling conclusion that real scientific research cannot flourish under a totalitarian regime, and here he names also Nazi Germany. This is rather surprising, coming from somebody who had access to everything done by the other side in the last war, and we now know that what the other side did was by no means negligible. The fact that the German regime mismanaged its whole scientific effort, and even botched the results of the scientists' work when they achieved something, should not detract from the quality of the scientific work that was allowed to proceed. However, although nearly half the book is devoted to this kind of critique, Bush's conclusions about future warfare are really the important and thought-provoking part, more so than his opinions about past and future adversaries, misguided as his thoughts, or adversaries, may be.

Bush's first conclusion was that war in its then-present form had passed from the world because of the appearance of jet aircraft, the proximity fuse, radar, and the atomic bomb.[19] Because of fighter and ground defenses, aircraft would not be able to attack ground targets.[20] Although high-flying aircraft would be able to use guided bombs, they would not be able to hit specific targets in a city.[21] Bush admitted that there were already television-guided bombs, but these verged on the warfare of Flash Gordon and Buck Rogers.[22] He discussed at length aircraft carriers and their protection against attacking aircraft armed with guided missiles and came to the conclusion that the future navy would have to operate only against enemy submarines. Consequently, "The day of the great ship may be over. Never again may the world witness the moving spectacle of a mighty ship, manned by two thousand men."[23] Because of their velocity and their huge turning radius, as much as a mile or two,[24] jet aircraft would be incapable of dogfighting and would serve only to attack bombers and transport aircraft.[25]

Future Warfare

So, conventional strategic bombardment by means of bomber aircraft would thus pass away, but long-range missiles were not to be feared either. Long-range missiles like the V-1 would be easy to shoot down by antiair artillery equipped with proximity fuses. (In truth they were easy to shoot down, but then they were first-generation, nonmaneuvering, flying straight and level. If television-guided bombs were indeed Buck Rogers stuff, then the Tomahawk would be at least Martian in origin.) As for ballistic missiles flying above the atmosphere, they are actually useless at long ranges. The V-2 carried a payload of one ton to a distance of about two hundred miles. It is true that range can be increased at the expense of the payload, but with a single-stage rocket, at about four hundred miles there will remain no payload to carry.[26] (Unfortunately, the North Koreans and the Iraqis proved different.) It is true, Bush continues, that a multistage missile can be built to get around this problem, but its price will run into millions. "If we are content to pay millions of dollars for a single shot at a distant target, it can be done in this way for any stated distance."[27]

But there are more problems. A missile traveling to a height of a thousand miles might burn itself up like a meteor on reentry, so the whole concept of attaining long range by rocket weapons is a pipe dream.[28] But even if this problem is solved, there remains the problem of accuracy. After a flight of about two hundred miles, the V-2 missed its targets by up to fifteen miles. If such a missile were fired to a range of two thousand miles it would miss by 150 miles. Even if the most advanced guidance systems (like in an aircraft) were applied, it would miss, in most cases, by up to ten miles, and if all went very well, within a mile.[29] Also, "We even have the exposition of missiles fired so fast that they leave the earth and proceed about it indefinitely as satellites, like the moon, for some vaguely specified military purpose. All sorts of prognostications of doom have been pulled from the Pandora's box of science, often by those whose scientific qualifications are a bit limited, and often in such vague and general terms that they are hard to fasten upon."[30] Admittedly, at that time space flight was not considered suitable even for the birds. But it seems incredible for a man of Bush's stature and experience, in 1949, to deride in such strong terms the concept of space flight and artificial satellites.

Although neither a high-flying bomber nor an intercontinental missile poses a major threat, a surprise attack can still be carried out by a short-range missile fired from a submarine. On this subject, Bush, a scientist who is also well versed in political life, makes a strange political statement, which has nothing to do with science but with doomsday national strategy. In such an event, however, "We in the United States have to defend only a relatively few points when it comes to attack from the sea by submarine. We would not defend Miami, for example, much as the people of Miami might object, for the activities of Miami are not essential to the prosecution of a war effort, and the question would be one of national survival when the issue became joined."[31]

Yet atomic weapons, after all, are not such a terrible threat either. There is a shortage of uranium, and this will make bomb production very expensive. Thus it will be perhaps quite a few years until both sides will frown at each other from behind great piles of atomic bombs.[32] How inaccurate all these predictions were can be seen from the following table:

Table 6-1: Number of Atomic Weapons Possessed by Both Big Powers, by Year[33]

Year	*U.S.A.*	*S.U.*
1945	2	—
1947	13	—
1949	170	1
1951	438	25
1953	1,169	120
1955	2,422	200
1957	5,543	650

Bush concedes that the ability to obtain enough atomic bombs depends on the capabilities of the adversaries of the United States, but a totalitarian regime will not have the necessary plants for sophisticated products, nor the manpower with the required skills and the resourcefulness of free men.[34] The facts are somewhat different, and in his position Bush should have known them. After the start of the war, the Soviet Union relocated its industries east and restarted production under extremely harsh conditions and in record time.[35] On the other side of the fence the situation was similar. In many fields of military science the Germans showed great resilience and inventiveness. In spite of the Allied bombing campaign, German total armaments production rose over 300 percent between January 1942 and July 1944, and as late as November 1944 it stood at 260 percent of the January 1942 levels.[36] In many instances the Germans failed because of political intervention and because by the time they got around to do things correctly, they were handicapped by the general military situation and the lack of resources. Nothing was wrong with their inherent capabilities. But Bush goes on. There is no doubt in his mind that the Nazis suffered from the same problem. They after all did not develop the proximity fuse, because "the techniques of this fuse were just too much for their regimented science."[37] However, a few pages earlier Bush states, in reference to the V-2, that "it was an expensive and complex affair, involving great ingenuity and remarkable engineering skill."[38] As for radar and electronics, it should be pointed out that except for their initial failure in the centimeter-radar field, German radar work was considered quite well advanced.[39]

Bush predicted also that there would be no way to develop an artillery shell carrying an atomic warhead.[40] Such a shell was proposed in 1949 and the first one fired in May 1953. Product development of successive generations of such shells lasted into the eighties.[41] Bush predicted that atomic energy could probably serve to power large ships, but they were of waning importance in war, and in any case the United States would have no prospective maritime enemy.[42] Finally, Bush discussed several ideas for which he did not have the competence or simply got confused. He claimed that jet aircraft would not be able to conduct fights based on human vision alone without radars,[43] and apparently he was hazy about jet engine terminology.[44] Finally, he claimed that the V-2 was radio guided from the ground.[45] This was true only for some test flights. The operational missiles all had inertial guidance.

Bush's book can serve as entertaining reading today, but in line with the theme of the present volume, his text contains some very important lessons on the pitfalls of completely unfounded scientific shortsightedness, doubly so when coming from a person with Bush's eminence. As can be seen, the problem with Bush's views consists of two parts. The first is his contention that nondemocratic regimes cannot produce real science or technologically advanced work. The second is his predictions about future wars. These two factors lead to some very disturbing conclusions.

German and Soviet Science

There is no question that in order to approve American research proposals, Bush himself, as well as the members of the NDRC and the OSRD, had to understand these proposals and their relevance to the ongoing war. This means they also had to have some idea of what the other side was doing. They had to know—nay, they knew—that in certain fields the Germans were ahead of the Allies. Robert Buderi describes how in the early part of 1943, Bush was "stunned" at the German U-boat successes, and he thought that the Allies were headed for catastrophe.[46] The Germans were the first to develop a jet fighter. They failed to use it properly, not because their scientists were unable to put two and two together, but because of an idea-fix of a half-mad leader (see chapter 7). The Germans also developed rocket-powered interceptors and used them operationally. Their advances in the science of aerodynamics were extraordinary—they were the first to develop the concept of the swept-back wing, necessary for high-speed flight, which now is a standard feature on almost all high-speed jet aircraft, including passenger planes. As a former head of NACA, Bush in particular should have appreciated the novelty and importance of this concept. Admittedly, during the war the Germans fell behind in the development of new radars, but on the other hand, theirs were of higher quality. A British scientist once commented that it was easy to distinguish between a German radar and a British one: the German radar's broadcast was more stable.[47] The Germans were the first to introduce electronic navigation aids for night bombing, and they were the first to employ guided missiles for antishipping strikes.

The discovery that the Germans were developing both cruise missiles (V-1) and ballistic missiles (V-2) came as a total surprise to the British, especially considering the huge effort the Germans put into these projects. The V-1 project was discovered only after the Germans started building the launching ramps that could not be hidden. Furthermore, there were dozens of smaller developments and inventions that did not have a potential of being "war winners" but were quite useful in their small way. The Germans did lose the war, partially for not properly managing some scientific developments, but mostly because of an inept political leadership. But from there to the claim that their science was deficient because of the regime under which it operated was a very long leap. So why did Bush write and publish such claims?

The most likely possibility is that Bush wrote what he actually believed in, having let his scientific objectivity be clouded by personal feelings. After the war, and particularly after the high-handed, and often ungracious, behavior of the Soviet regime toward the western Allies, and coupled with various aspects of the conduct of the Red Army, the west formed some very poor opinions about the Soviet Union. The prevalent belief was that the Soviets were a bunch of uncouth *muzhiks* (Russian, meaning an uneducated farmer of the lowest grade) who could barely read or write. Except for the fact that the regime had at its disposal several hundred divisions of savage Mongols, there was really nothing to worry about from them. They were technologically backward, and the few things they did develop were either stolen or copied from here and there. Was Bush only the herald of these ideas or actually one of their originators? There is no doubt that Bush was not the only person who held such beliefs.[48]

In August 1957 the Soviets announced that they had developed a long-range rocket system that could take an atomic bomb anywhere on the face of the earth. The western world had a good laugh at the expense of the lying Soviets, although for years the United States intelligence community was reporting that the Soviet Union was working on long-range rockets.[49] On October 4 of the same year, the Soviet Union launched the first Sputnik, and the laughter changed into shock. Luckily for the rest of the world, even the Americans probably did not take Bush's predictions seriously, because they geared up and in twelve short years beat the Soviets to the Moon in the space race. On the other hand, the west (i.e., the United States) did lag for several years behind the Soviet Union in large rockets, and there was for a while some talk about a "missile gap."[50] Did Bush's opinions carry enough weight to influence American defense thinkers into complacency at least? Perhaps making generalizations about the capabilities of various regimes is not such a good idea.

The Danger of Preconceived Ideas

The second part of the problem consists of Bush's completely unsubstantiated (and very soon to be proven completely wrong) predictions about the future of various military technologies. What is worse is the fact that much of the theory to prove him wrong about several of these predictions already existed and was available at the time he wrote his book. The question of financing some of these developments, like the multistaged missiles, appears to be a legitimate one. But who was better qualified than Bush, who during the war "spent" billions, to know that when the chips are down money is a secondary consideration? Actually Bush, who understood operations research, should have known that the real question was not money, which can be borrowed or even printed when necessary, but resources, manpower, and materials. So how did Bush, a distinguished scientist with a proven track record, go wrong, and worse, prove so conclusively wrong in so short a time? Practically every prediction he made, from the capabilities of jet aircraft, through the proliferation of nuclear weapons, to ballistic missiles and space flight, was proven wrong in less than ten years.

The first possibility that comes to mind is that Bush was asked to write such ideas as part of an effort in intelligence warfare. Somebody was trying to feed the Soviet Union some mistaken notions about American thinking on future weapons development, presumably to channel the Soviets' efforts in this or that direction. And who better to give it an air of respectability than the former head of the OSRD? Bush is shown here in a very unflattering way, but maybe he was willing to swallow this pill for patriotic reasons. Although it is possible that this kind of harebrained idea was proposed (there were worse), it should have been clear to everybody, including Bush himself, that such a stratagem would work for no more than a few months. Considering the ongoing research efforts in many fields (even with the customarily shrinking budgets after a war), the Soviets, who were sure to follow American policies and technological developments, would have quickly seen through the ruse.

The other, less pleasant possibility is that Bush actually believed in what he wrote. One cannot presume to guess by what thought process he came to these conclusions, but somehow they were arrived at and published. One thing, though, we can be sure of. This kind of thought pattern does not develop in days, or even months! Did Bush, in his capacity as the head of the OSRD, judge all novel ideas the same way he judged ballistic missiles and future atomic weapons? How many good concepts were brushed aside because Bush thought they were fantasy? Bush actually hints in his book about his (hidden?) approach to the question of innovation. "The notion that all things are possible to the scientist is amazing, and it produces foolish statements."[51] This statement raises another major problem. Who is to decide, a priori, what is possible and what is not? The current head of the R&D (research and development) effort, a committee, a Nobel laureate, or maybe a Delphi poll?[52] The possible answers to these questions are very disturbing.

Of course, not everything is possible, not even to scientists, but experience has proven—and is proving constantly—that given adequate financial support, most of the ideas promoted by scientists are feasible. Bush of course was not required to foresee and propose all such innovations by himself, but he should have been more careful in judging them useless. It can be legitimately argued which innovations are necessary, or even are worth the expense of development (as if someone can define worth in such circumstances), but this is essentially a political decision, even in the armed forces, and not a scientific one. The reason so much space has been devoted to Dr. Bush and his book is to point to a danger that exists even today. When an "old man of science" with impeccable credentials and a solid record of achievements, in any scientific field, expresses an opinion, or a set of opinions, in what ostensibly is his field of expertise, who can argue differently, or for that matter say "the king is naked"?[53]

Bush made a cardinal error, admittedly common but still inexcusable in a scientist. He photographed in his mind, so to speak, a certain state of technological and scientific development. But to continue the analogy, this was a "still" photograph, and he was totally blind to the fact that both science and technology are in a permanent fluid state, moving constantly forward. Bush (presumably) saw this during the war, and with a little effort should have recognized this trend in action during his own career. Not realizing these changing circumstances was really the root of his mistaken predictions.

7 Political and Ideological Meddling

Political leaders, whether civilian or military, had more influence in the outcome of the World Wars and the shape of policy in the interwar years than the military command. It was they who carried their countries to victory or doomed them to defeat.

Allan R. Millet and Williamson Murray

INTRODUCTION

It is generally accepted that the political leadership of the state formulates policy, be it economic, political, or military, but leaves the execution of this policy to the professionals. The sight of the head of state leading his armies in the field passed away with Napoleon. One of the reasons for such an approach is that running the various functions of a modern country is extremely involved and thus is to be apportioned to the various specialists.

Although it seems that this is the best way of running things, there are still two potential sources of trouble in such an arrangement. One occurs when the appointed military commanders try to usurp the political decision making. In the military field this could be disastrous. Two good examples are the Japanese involvement in Manchuria (1931) and in China (1937), both of which were brought about by the army and forced the government's hand. The China operation drew world attention and developed into a major drain on Japanese resources, with no real benefit to the Japanese. The other side of the coin is when the politicians start getting involved in purely professional decisions. Today this kind of interest is becoming even worse, particularly in the military field. It was bad during World War II, where Hitler for example intruded in purely military matters. It was worse during the Vietnam War, with "micromanagement" raising its head. Today, with worldwide instant communications, the "decision makers" in the rear can interfere down to the level of the squad leader. The politicians often think they know better than the professionals. This may be perfectly true, but the best way to manage is to fire the suspect professionals and appoint others. However, the source for the worst failures involves mixing abstract ideological considerations into practical reasoning. In the military field it can amount to demands for results the military cannot achieve, either for lack of resources (usually the political arm's fault) or for other, more abstract reasons. Still, while occasionally the military can compensate for unreasonable demands by physical sacrifices, such demands in the scientific or technological fields always prove impossible to carry out. The laws of nature, after all, are considerably more formidable than the toughest politician or general.

Although this kind of behavior can happen in democracies too, especially when there is competition for resources, senseless political and ideological meddling is more common in dictatorships. Democracies are considerably more flexible on points of ideology, and they usually understand that ideological, and thus nonfunctional, considerations are detrimental to pragmatic interests. Probably the most glaring example of the truth of this statement, and with the worst consequences, was Nazi Germany and its treatment of its own scientists.

NAZI GERMANY AND SCIENCE

German Rearmament

From the beginning of the First World War, it was evident that the might of the armed forces was becoming more and more dependent on the men (and women) of science and technology. It was aptly demonstrated in practically all fields of warfare. Between the two world wars this fact should have still been vivid in the recollection and understanding of all concerned. In other words, human resources of any kind, but particularly the scientific ones, are too precious to be squandered in any way.

When the Nazis came to power in Germany, they did two things that were in line with their avowed preaching. In order to erase the shame of the defeat in the previous war, Germany unveiled and enormously strengthened the ongoing German rearmament program, which was initially started by the Weimar Republic in the early twenties. In April 1922 (long before Hitler), Germany and the Soviet Union signed the Rapallo Treaty, which regulated the relations between the two countries, while waiving both sides' demands for war reparations. The treaty also had a secret component, about establishing centers for development and cooperation in three fields of military science—aviation, armor, and gas warfare. This cooperation ended when the Nazis came to power.

What's more, the Nazi regime carried out this rearmament program on a much grander scale, producing weapons in mass quantities. Surprisingly enough, this program was not always as innovative as might have been expected from a scientific and technological power such as Germany. As pointed out above, after touring Germany in 1937, Alexander de Seversky claimed that when everything was said and done, the Germans relied more on quantity than on technical/operational thinking. Events bore him out. While in several fields German doctrinal thinking was without doubt very advanced, in many other cases they did not think things through far enough. Consequently, in many respects the Germans painted themselves, technologically and scientifically speaking, into a corner, from which they failed to extricate themselves. Furthermore, they apparently did not realize that modern warfare is also a question of standardization, logistics, and carefully nurtured modern science.[1] Once the doctrinal innovations wore off, either because the Allies adapted to the existing doctrine or devised better technological systems to overcome it, the war became a war of resources. This was understood by everyone, but only a few people realized that resources were no longer measured only by oil and steel but also by science. Somehow, in Germany there were not enough ongoing scientific and technological advances to support the armed forces, let alone to create and support a new and improved fighting doctrine. Moreover, there was no coherent management to organize what scientific work was being carried out. While several new weapons were of course quite impressive and innovative, they were, so to speak, peaks and spikes and not a consistent scientific thinking combined with military thinking. In addition, the scientific advances were not enough. They either came too late or were too local in their effect. The best proof of this is the fact that the Germans became aware of the concept of operational research only in 1943. The German effort and what it achieved was still impressive enough to frighten some of the European leaders, especially when compared with what the western democracies did until then. But it can probably be rigorously proved that the initial overwhelming superiority of the German armed forces was more due to the fact that they started their preparations for war earlier, and not because they were better (the Germans did not expect a war for a few years yet and this would have given them time to better absorb and integrate also the industries of Czechoslovakia and Poland). They thus gained both the quantities of war materiel and the time to absorb them properly even before the war started. This gave the Germans a marked edge during the initial campaigns and enabled them to lay their hands on the economic and industrial resources of all of central and western Europe as well. Thus they had a considerable leverage, which almost brought them victories in several fields. However, once the initial thrust was contained, Germany could not repeat the previous successes and slowly ran out of steam.[2] The question, though, remains: How did Germany, with by far the highest number of Nobel laureates in the sciences find themselves in such an inferior position?[3] To find the answer, we must look again at some of the doings of the Nazis.

Nazi Cleansing of the Scientists

An important aspect of the Nazi credo was the cleansing of the land and the people of the Reich. Consequently, "nondesirable elements," mostly Jews, were persecuted and where applicable, fired from all state and other public institutions. This also included a lot of perfectly good Gentiles who happened to have a Jewish spouse or Jewish ancestry. But the problem in Germany went considerably deeper than just mistreating a group of Jewish intellectuals who were kicked out of their jobs. The German people had a long and honorable tradition of respect for higher education and science, which was probably more deep-rooted and more pervasive than in most other European countries, and attested to by the Nobel statistics (even in literature, until 1939, Germans had won five of thirty-eight awards, second only to the French, who had won six). But apart from the few German intellectuals who truly embraced the Nazi dogma, most of the rank and file of the Nazi party in Germany were perennial misfits, thugs, or party hacks, who rose to power with the Nazi party. They did not appreciate the contribution of the intellectual elite to the country and its people, and under their rule the title of "professor" was on the way to becoming a derogatory term. The result was that apart from many real or suspected Jews who left their positions and eventually Germany, many non-Jewish intellectuals emigrated from Germany. One of the more famous ones was Erwin Shroedinger (actually a Catholic Austrian), the Nobel laureate in physics for 1933, who left Germany that same year. Ironically, in a newspaper interview in 1931, Hitler was asked about the potential damage that his avowed policies toward the Jews would cause if he ever got to power. Hitler dismissed the Jewish contribution as inconsequential.[4]

Modern science does not operate in a vacuum. Compared with the scientific giants of old, today's scientists constitute part of an "establishment" to which they contribute by research and teaching and from which they absorb intellectual sustenance. In years past this scientific establishment was very cosmopolitan in nature, and the scientists generally did not get involved in the "petty bickering" or even wars between their respective countries. Liddell Hart noted that even during the Napoleonic Wars, British scientists traveled to the continent and were not molested. This freedom of action was curtailed during subsequent wars because of the demands of secrecy, but even during the height of the Cold War, there was a lively exchange of information on topics not deemed of military importance.[5]

A healthy scientific establishment can lose a certain number of active scientists and still continue to function. After all, in any large country hundreds if not thousands of scientists retire each year and scientific work does not come to a standstill. Such a scientific establishment can even survive the sudden accidental demise of active scientists or even key people. But the wholesale sudden eviction of scientists, educators, and professional people, in all fields of science, technology, and the arts would create empty voids that could not be hoped to be filled for many years.

Germany's entrance into the war was, from her point of view, highly premature. Germany was preparing for a long struggle against the democracies and the Soviet Union and was in the middle of a huge arms buildup. The German leadership did not expect, nor was it really prepared, to be involved in a shooting war for another four to five years. When he invaded Poland, Hitler did not expect France and England to honor their pact with Poland and was genuinely surprised by their declaration of war against Germany. In the present case, however, within a few years of the Nazi ascendance to power, Germany found itself involved in a war (based almost entirely on modern technology) and with the whole military system crying out for a strong science.

This problem was compounded for Germany by three more aspects that apparently it did not fully consider. The exiled scientists did not evaporate into thin air. Most of those who managed to leave Germany (some who could foresee what might happen even before the Nazis came to power) found asylum in other countries and eventually went to work for the Allied war effort. The most important of these was Albert Einstein, who moved to the United States. Another was Leo Szillard, who moved to England and was later instrumental in starting the Manhattan Project for the development of the first atomic bomb. Another one was Edward Teller, who moved to the United States and is considered the "father" of the hydrogen bomb. Still another was Theodor von Karman, an aeronautical researcher who contributed enormously in this field to the Allied war effort. And there were scores of other scientists who were able to leave Germany before the war. This story repeated itself after the war started and the conquest of the rest of the continent. Many other Jewish scientists who were not thrown into concentration camps managed to escape and ended up also on the Allied side. There were initial difficulties with some of these people coming from "enemy" countries, but in most cases these problems were eventually solved.

In all fairness it has to be said that the Nazis did not mistreat only the Jewish scientists. They mistreated their own too, but in a different manner. The overwhelming successes in the initial stages of the war convinced the Germans that they did not need a real scientific effort in order to continue the struggle against the British. Consequently, research projects were stopped or slowed, teams of scientists were disbanded, and many of them were drafted into the armed forces. General Josef Kammhuber (commander in chief of German night fighters) had to plead with Hitler to allow him to continue development of airborne radar, which was just ready in early 1942, when the British night bombing became more than a mere nuisance.[6] Furthermore, because of the draft, the orderly studies of many students were disrupted. In various fields this affected whole graduating classes, who went into military service instead of graduate studies and research and left in place only the older generation to continue the work. Although there is no doubt that this older generation had the experience, the situation had another subtle effect on the quality of science. It is generally accepted today that the most fruitful years of scientists, particularly in the fields of physics and mathematics, are when they are in their early twenties. But this was exactly the age group snatched from their studies. It is interesting to speculate whether the Germans' failure in the atomic field and their very late entry into the field of operations research were not affected by the draft. In other words, the total human professional base, absolutely necessary for a long-range scientific effort, was badly shaken, with all attendant consequences.[7] Admittedly, when in the winter of 1941–1942 the Germans realized that the war was not going to end soon, they turned about and started rebuilding their scientific establishment, but it proved too late because nearly two years had been lost. In a way this was similar to the military collapse of the Allies in May 1940, caused by about a two-year lag in rearmament and formulation of appropriate doctrine.

Because of the accumulation of all these factors, the Germans in essence put more emphasis on engineers and less on scientists. While engineering is important, there is a definite division of work between technicians, engineers, and scientists. Chapter 5 (on the American torpedoes) of this text notes the comments about giving scientific work to overgrown draftsmen. These problems were multiplied many times in Germany. The Japanese were even worse in this respect. Many of their leading scientists were trained in the west and were not trusted.[8]

Finally, the Nazis blundered in one more way, and this was the most bizarre. In Nazi Germany, physics was considered a "Jewish Science," and naturally this made it immediately suspect. It is true that there were many Jews in this field, starting with Einstein and ending with Lise Meitner, Otto Hahn's assistant in 1938 when he achieved intentional fission of the uranium nucleus, but assigning one of the natural sciences to a racial group was truly the height of folly. The Germans, however, stuck to their guns. Physics students never received deferment from the draft, and in Germany, research in nuclear physics was under the direction of the postmaster general.

Of course, some noted physics researchers did remain in Germany. One was the above-mentioned Hahn. Another was Werner Heisenberg, the 1932 Nobel laureate in physics, who is known for his statement of the "Heisenberg Principle."[9] But physics research in Germany remained moribund. This resulted from a lack of intercourse with capable colleagues (many of whom left), the suddenly smaller size of the German physics community, the lack of contact with the scientific community abroad (caused by both the war and the abhorrence of anything German), and the disdainful attitude of the German authorities. This decline in physics research was particularly apparent when compared to huge advances made later in Germany in other fields such as aeronautical research and rocketry.

While not dealing with science, it is interesting to note that

Nazi ideology had another effect on the German war effort. The Nazis never fully utilized women to take the place of the men who were mobilized for the war effort, as was done in England and particularly in the United States. The Nazi ideology held that the woman's place was at home, raising more little Nazis. When the industrial manpower shortage was realized, the Germans started using forced labor from the occupied countries and from concentration camps, but this could never be as efficient as using their own people.

THE WASTE OF THE Me-262

While the story of German science is one that was true essentially of the whole country and the regime that applied the ideology, there were cases where the will of a single person clinging to outdated illusions that had nothing to do with the Nazi ideology caused further irreparable damage. As head of state and supreme commander of the armed forces, Hitler is notorious for the way in which he intervened in the day-to-day running of the military, by way of what is today referred to as micromanagement. Many of Hitler's strategic and operational directives were calamitous, like his insistence not to retreat voluntarily on the eastern front or his orders to General Friedrich Paulus not to attempt a breakout from Stalingrad when it was still feasible. These two errors were caused by Hitler's belief that no matter what, conquered territory should not be given up. Corollary to this approach was his constant harping on the need to attack, and his neglect and delay in approving defensive measures against Allied bombing, contrary to the advice of the professionals.[10] Throughout the war Hitler insisted that bomber production had priority over fighter production, even as Germany was being reduced to rubble under the combined attacks of Bomber Command and the Eighth Air Force. The failure of the first operational jet fighter in the world to enter into combat in an orderly and effective way was the culmination of this irrational behavior.

Like the previously described development of the radar, so it was with jet engine technology as a prime mover for aircraft, particularly fighter planes. Several countries, notably Britain, Germany, and Italy, were pursuing the development of these novel power plants. The jet engine promised several important advantages over piston-engine equipped, propeller-driven aircraft. These advantages included greater reliability, smoother running with less vibration, and simpler and apparently unlimited ability to convert mechanical power to propulsive force. Propellers are limited in this respect by their tips' speed, which cannot exceed certain velocities. The basic ideas about jet propulsion were established in the late twenties, and the theory was well in hand in the beginning of the thirties. The first patent for a jet engine was applied for in 1930, in Britain, by Frank Whittle, then a test pilot and flight instructor in the RAF. Another one was registered in Germany in 1932. The only problem was that the necessary technology, particularly in the field of metallurgy, was lagging behind the theory. But the lack of understanding of the whole concept of propellerless engines and their advantages was so pervasive that in 1934 the British undersecretary of state for air released the following statement: "We follow with interest any work that is being done in other countries on jet propulsion, but scientific investigation into the possibilities has given no indication that this method can be a serious competitor to the airscrew-engine combination. We do not consider that we should be justified in spending any time or money on it ourselves."[11] This concern for public monies was probably the reason that the Air Ministry did not approve the payment of the five-pound fee (about twenty-five dollars at the time) required to renew the original British patents. Frank Whittle, in whose name the patent was issued, was unable to afford the payment out of his pocket and so the patent lapsed.

Slowly, however, the researchers overcame the technological problems. Finally in Britain the subject was recognized as being of importance, and the British government embarked on an orderly development program of jet engines and suitable aircraft. The first British jet flight was made in May 1941 in the Gloster E-28. The Americans were initially informed of the jet engine development by the Tizard mission but did not show enthusiasm. In April 1941,

General H. H. Arnold, chief of staff of the Army Air Force visited England and saw the British jet taxi and hop into the air. He was enormously impressed[12] (in America the concept had not advanced beyond the drawing-board stage) and arranged for an active British-American cooperation in this field.[13]

In Germany, on the other hand, the situation was somewhat confused. The earlier attempts to employ a new mode of propulsion were made by Ernst Heinkel, the well-known aircraft producer. After a meeting with Wernher von Braun, the leader of German rocketry, a small existing airframe (He-112) was adapted for this purpose, and after many difficulties and failures, this rocket-powered plane achieved a successful flight in the summer of 1937. Heinkel then embarked on a more ambitious project, using a different rocket motor and a specially designed airframe, the He-176. This one achieved first flight on June 20, 1939. The next morning it was successfully demonstrated to the German aviation leaders, but they were angry with Heinkel for developing the airplane without the Air Ministry's sanction (and for previous other actions by Heinkel, whom they deemed too independent) and actually banned any further work on it. Heinkel persisted and in July arranged a demonstration before Hitler and Goering, but they too considered the plane a useless toy and no action was taken to push its development. When the war started in September, the He-176 was actually sent to a museum. Heinkel, however, was a man of many talents. Starting already in 1936 he privately financed the development of a turbojet engine under the leadership of Pabst von Ohain. Although this project was not financed with any state funds, the Air Ministry told Heinkel to leave engine development to engine manufacturers and concentrate on his specialty—airframes. The engine manufacturers on their part were too busy building piston engines for the growing Luftwaffe and did not want to get into new ventures. However, even the Air Ministry could not completely ignore the facts of life, so in October of 1939 a contract for the development of a turbojet engine was awarded to Junkers and BMW, and an airframe contract was given to Messerschmitt.

The Air Ministry thus got back at Heinkel, but this was doubly strange. Even before these contracts were awarded, the new He-178, with the Ohain jet engine, flew successfully for the first time on August 27, 1939, with the same pilot who rode the rocket planes two months earlier at the controls. Air Ministry officials were present, but Goering himself did not bother to come. He probably had made up his mind by then to bypass Heinkel. However, the overwhelming success of the Polish campaign, and the expectation that pretty soon the British would come to their senses, brought Hitler to order a stop to all long-range research projects, defined as those that would not enter production within a year. This policy in fact was maintained until the end of 1941 when, after the failure to take Moscow, it became clear that no quick victory in this war was forthcoming.

The German aviation industry, however, paid little heed to this order. Messerschmitt continued the construction of his Me-262 (for which he still did not have an engine) and even had a rocket plane (Me-163) designed and built. In May 1941 this rocket airplane broke the world's speed record by reaching one thousand kilometers per hour, although because the Germans did not want to disclose the airplane's capabilities, it was never announced as such.[14] Heinkel too continued the development of a new jet airplane, the He-280, which made its maiden flight on April 2, 1941. On April 4 the first Me-262 flew with a piston engine; the BMW engines were not ready and the Junkers' development was even further behind. In March 1942 the Me-262 flew with the newly arrived BMWs, but with the piston engine still in place in the nose. This proved to be a wise precaution, because both jet engines failed immediately after takeoff. Finally, in July 1942 the Me-262 made a successful first flight with the Junkers engine. There was one problem though. The undercarriage was the standard one for practically all aircraft of that period—two main wheels in the front and a tail wheel. Due to the lack of propeller airflow (there was no propeller), it was almost impossible to raise the tail to get it into the airstream and get longitudinal (nose up and down) control. Pilots had to revert to a dangerous maneuver with the wheel brakes to achieve takeoff. The chief test pilot wanted to have a tricycle gear installed, but the Air Ministry refused to adapt an "American invention."[15] The gear was finally adapted, but only after Adolf Galland, inspector general of fighter aircraft, flew it in May 1943. It seems that at this stage Heinkel gave up the efforts to introduce his airplane and went back to building airframes.

Meanwhile, back in Berlin, Hitler was not apprised of this development, partially because work was continued during his ban on development projects. Now, however, several factors combined and made such a revelation necessary. First, Galland wrote to Goering and Erhard Milch (at the time director of air armament) an enthusiastic report on his flight that they could not ignore. Second, the round-the-clock bombing of Germany was starting to be felt, with the greater danger poised by the increasing American daylight effort. In the meantime, the German fighters were barely holding their own against the Eighth Air Force, but initial longer-range excursions by Thunderbolts equipped with drop tanks were already recorded. The German leadership understood that now was the last chance to clear the skies over Germany, and the Me-262 was an ideal candidate for this role. Finally, even at this stage, Hitler did not want fighters for defense but bombers for attack. Production of a new fighter would have to be approved by Hitler. So after a delay of several months, on November 26, 1943 (when the P-51 Mustangs had already started flying over Germany), the new fighter was demonstrated to Hitler. After the demonstration, Hitler turned to Willi Messerschmitt and asked him if the new airplane could carry bombs. Messerschmit, an astute engineer if ever there was one, answered, "Yes, any airplane can, but . . ." Hitler did not let him finish and declared, "Here at last is our Blitz bomber."[16]

Hitler was absolutely consumed with thoughts of revenge against England, which could not be bombed successfully. A quick dash by a fast bomber might turn the trick. Besides, Hitler decided that this new bomber would be used against the forthcoming invasion. But the technically minded people around him understood the problems better than Hitler did. First, the bomb load would rob the fighter of all its performance advantage. Second, there did not exist a suitable bombsight for horizontal bombing by a single-seat fighter. Dive-bombing was out of the question because in a dive this sleek and aerodynamically clean, the airplane would develop excessive speed and break up upon pulling out. This of course could be remedied by strengthening the structure, but again this meant weight and reduced performance. Finally, the high fuel consumption at low levels would necessitate so much fuel to be carried to reach England that no useful load could be carried. Some initial testing confirmed all these forebodings, but Hitler was adamant. So the Luftwaffe command, on its own, decided to continue to develop the fighter version but without consulting or informing Hitler.

In April 1944 a memorandum was presented to Hitler recommending that the aircraft industry devote all its resources to the production of jet fighters instead of piston-engined ones. Hitler asked how many bombers had been produced up to then, and when told there were none, he flew into a tantrum and forbade any mention of the airplane as a fighter or even a fighter-bomber. The fighter development was stopped, and the conversion of the fighter into a bomber dragged on. It was first used in August of 1944, long after the D-Day invasion, and of course nobody talked about bombing England. This was still essentially impossible with the Me-262. In September, Albert Speer, in charge of all German armaments, took it upon himself to return to develop the fighter without consulting Hitler. The bombers were also reverted to the fighter configuration. All these orders and counterorders created a good deal of disorder. The airplane was not used optimally, and furthermore the construction of a two-seat training version was not entered into until November 1944. This was an important point. By then Germany was running out of experienced pilots, who in combat with the Allied fighters were being decimated at an increasing rate. Furthermore, the fuel situation in Germany was beyond critical because of Allied bombing against the oil industry. This in turn curtailed pilot training to the extent that novices were being sent into combat after only a few tens of hours' training. While the Me-262 airplane in itself was easy to fly, many pilots crashed because they had difficulty in quickly adapting to an airplane with a completely different "feel" and engine-control regime.

There is no doubt that Hitler's emotional and illogical intervention in many matters caused Germany tremendous harm. Here, however, was a case of a political leader trying to change the laws of physics and aerodynamics. Had the Me-262 been put into an accelerated program of development and training toward the end of 1942, it could have been ready for combat at about the middle, or toward the end of 1943. This was prior to the Allies' wide-scale introduction of effective long-range fighters such as the P-47 and the P-51, which were piston-engine powered. The Allied jet developments centered on three airplanes, but they were all behind the German developments in this field. The British Gloster Meteor, which made its first flight only on March 5, 1943, entered squadron service in July 1944. The American Bell P-59 prototype first flew in October 1942, with a British engine, and the Lockheed P-80 in January 1944. Admittedly, several months could have been shaved from these programs, but it would have been a touch-and-go situation even then.

The Me-262 would have had a fairly free hand to stop the B-17 raids before any of these Allied planes were produced in meaningful numbers. This would have given the Luftwaffe time to recuperate, as happened with the British Fighter Command toward the end of the Battle of Britain, when the Germans shifted to bombing London. The Luftwaffe might even have regained control of the skies on the eastern front. The Normandy invasion likely would have had to be postponed on the very same grounds that Sea Lion (the planned German invasion of England in 1940) was postponed and later canceled: no air superiority.

How the war would have looked after that is anybody's guess. The atomic bombs might have been used against Germany first (this was contemplated by Vannevar Bush)[17] or the American and Soviet materiel resources would have still carried the day. In any case this is speculation. One thing is sure, that micromanagement of purely technical or military matters by the political leadership is useless and will lead to disasters.

HOW THE BOMB SAVED SOVIET NUCLEAR PHYSICS

Early Developments

As stated earlier, political and ideological meddling are not limited to democracies or dictatorships, nor are they limited to any particular color of dictatorship. The chain of events that will be described in the next section did not end in failure, but was very close to being so. What's more, it came on the heels of, and was fueled by, another ideological struggle in which reason lost and caused a long-term and embarrassing Soviet dependence on the west, as well as a drain on its resources, even during the Cold War.[18]

During World War II, through several well-placed spies, the Soviets received large amounts of timely and extremely accurate information about American progress in the development of atomic weapons. These spies even supplied copies of computations and drawings of components. Stalin was informed, in less than a week, of the first successful test of the A-bomb in Alamogordo, New Mexico, on July 16, 1945. When the war was ended with the dropping of the two bombs on Japan, the Soviets understood that strategic warfare had changed forever. Furthermore, they understood that if the Soviet Union was to survive, they too must have the bomb in the shortest possible time. There were in fact western generals who called for riding this success of American arms to its logical conclusion—to rid the world from Bolshevism.

The Russians established an atomic research institute near Moscow, and later when actual work on explosives (to be used to achieve implosion of plutonium cores) intensified, a second center was established in Sarov, in an uninhabited area some four hundred kilometers east of Moscow. Laurenti Beria, head of the NKVD (People's Commissariat of Internal Affairs), the forerunner of the KGB (Committee of State Security), was named the head of the project, (similar to the position of General Leslie Groves, who was the head of the Manhattan Project), and on these matters he reported directly to Stalin.

Party Intervention

In the Soviet Union, after the "great Patriotic War," Marxist ideology started to be interpreted in ways that Marx could not have envisioned. Furthermore, the ideological theoreticians of the Communist Party began to get involved in subjects where political ideology had no business whatsoever and of which its torchbearers had no understanding.[19] Consequently, actions were often sanctioned even if they contradicted logic or previous experience. One of the general targets of attack was the uncritical acceptance of the "western approach" to science and the low status of "Soviet Science" in comparison with western science, as presented in Soviet books.

One of the notorious examples of this approach, which eventually became a byword for the Soviet governmental gullibility, was the case of Trophim Lysenko. Lysenko had training in agronomy and botany, but first and foremost he was a dedicated Communist. He developed some "original" botanical theories of his own in which he repudiated the importance of chromosomes, claimed that environmental effects are assimilated and transmitted to the next generation, and made a hash of all theories of genetics and heredity. He was also a slick talker and convinced Stalin that if given a free hand, he would improve the state of the Soviet agriculture. This argument appealed to Stalin, and Lysenko was nominated to be the head of the Institute of Genetics in the Soviet Academy of Science. He promptly had all his scientific adversaries (those who practiced conventional biology) removed and ordered the termination of instruction of crucial parts of agriculture and biology. In 1948 he won a major victory. In the annual Soviet Congress on Biology, his theories on biology were accepted and the teaching of the role of chromosomes was banned in the Soviet Union. This in itself would not have been disastrous, of course, but it was only the tip of the iceberg. In reality, Lysenko's influence went much deeper. Within a few years, the Soviet agriculture was almost ruined and rather than being an exporter of grain, the Soviet Union became a heavy importer.[20] Lysenko was finally demoted in 1965, when Soviet science surged forward during the space race and Khrushchev attempted to beat the United States in every scientific field.

The nuclear research community in the Soviet Union was attacked in a similar way, particularly for adopting "defective western theories" about quantum physics and the theory of relativity. This last had already been criticized in the Soviet Union in the thirties, on philosophical grounds. According to the orthodox Communists, these were cosmopolitan (a euphemism for Jewish) theories and as such were not in line with Marxism-Leninism. With the progress and growth of nuclear research in the Soviet Union, all sorts of party hacks and political commissars started bothering the researchers, a high percentage of whom were indeed Jews. Stalin and Beria became worried. If Soviet biologists actually tried to sabotage Soviet biology, as Lysenko (whose opinions at the time constituted the official party line) claimed, did it not stand to reason that Jewish scientists would try to sabotage the Soviet nuclear research?

A new squabble broke out in 1947 when a senior Russian researcher published a paper in a Soviet publication, in which he affirmed some scientific theory advanced by Niels Bohr, a Dane, one of the most important researchers in the nuclear field and a Nobel laureate in physics for 1922. The Russian's article was bashed by a Communist philosopher, the editor of the publication was fired, and the "Danish approach" to quantum physics was also banned in the Soviet Union.

Lysenko's success in the 1948 biology congress encouraged the antiscientists, and within four months they started preparations for a similar congress on nuclear physics. In a letter to Klimenti Voroshilov, the deputy premier of the Soviet Union, the organizer explained that Soviet physicists did not receive the recognition they deserved, the textbooks were all full of foreign names, and the teaching of physics was done without reference to Marxist dialectic materialism. In the Soviet Union, this last charge amounted to heresy and was invariably treated accordingly. The last meeting of the congress' organizing committee took place on March 16, 1949. The congress itself was to start on March 21, yet sometime between these two dates it was canceled. The last-minute cancelation of a scientific congress is a rather unusual occurrence anywhere in the world, and doubly so in the Soviet Union, where from the start the whole event was sanctioned by the state. What happened between these two dates?

The Reasoning of the Soviet Leadership

In hindsight it becomes clear that for all of the Communist dogma, the Soviet leadership were practical people. Stalin felt the conference might retard the atomic project. Beria approached Igor Kurchatov, a longtime physicist and the head of the Soviet atomic project, and asked him "whether it was true that quantum mechanics and relativity theory were idealist, in the sense of anti-materialist."[21] Kurchatov answered that if these theories were to be rejected, the bomb would have to be rejected too. Beria took this answer to Stalin, who ordered the congress canceled while allegedly commenting to Beria: "Leave them in peace. We can always shoot them later."[22] The success of the first Soviet atomic explosion reestablished the standing of the scientists and canceled the influence of the "party line" on nuclear physics in the Soviet Union.[23]

It is interesting to compare Stalin's reaction to these two crises, the agricultural and the nuclear. The Soviet regime from the start was highly respectful of industrialization and thus of science and technology. The Five-Year Plans had addressed these issues, and the Soviet system put much emphasis on inculcating technology into the educational system and the general population. But the Soviet regime was not technocratic and did not consider the scientists immune to the party doctrine. The Soviet mistake was in allowing political theorists to interpret natural sciences by means of completely alien (Marxist) concepts. Stalin probably understood soon enough that Lysenko was a charlatan and that his theories were damaging the Soviet Union, but a few million starving Russians bothered him less than the importance of maintaining the supremacy of Marxist theory and the party rule. Nuclear physics was another matter. This was not a question of several million dead farmers. Here the future of the Soviet Union (and its function as power base for Marxism) was at stake, and a failure could very well mean capitulation to the hated west. In this case, practical considerations came first, and Stalin was willing to bend Marxist theory to that end.

8 Technological Developments and Science Fiction

No new weapon can be introduced without changing conditions, and every change in conditions will demand a modification in the application of the principles of war.

J. F. C. Fuller

Give them the third best to go on with; the second best comes too late, the best never comes.

Sir Robert Watson-Watt

NEW TECHNOLOGIES AND NEW WEAPONS

Introduction

It is generally accepted that during "peacetime" it takes several years—anywhere from five to fifteen—to develop a new weapon (or weapon system) from initial conceptualization to maturity, and this is for those that are successful and that eventually get into the hands of the troops. It can also be said that for almost every successful project there was another development project that for a variety of reasons—usually poor management, which resulted in cost overruns and schedule slippages—was terminated somewhere along the way. It does not matter if the weapon system we are talking about is an original concept that is intended to revolutionize the battlefield, or if it constitutes an answer to something that the other side has, and with which he intends to revolutionize that same battlefield. This time span of five to ten years is true when the technology that lies at the base of the new weapon is well at hand and no "unknown unknowns" occur in the process of development.[1] Furthermore, it is resignedly accepted that if a new technology is involved, the time periods we are talking about will increase by at least another five years.

The idea that long development times are a law of nature is mistaken, because it is based on poor assumptions and worse practices. Furthermore, the idea is usually a source for technological failures. It will later be shown that extremely complex systems were developed in a fraction of such time. The point, however, is that within these long time intervals there hides another, more subtle problem. If the development cycle takes up to fifteen years, or more, then it is accepted that the system under consideration will be fielded in fifteen years. This is not a great revelation. But then what kind of battlefield are we talking about, and what kind of variables might enter that might completely change the battlefield and render the present developments useless, or at least short of expectations? There is another difficulty with the "five- to fifteen-year rule." Technological innovations do not come in ten-year steps. While the gestation period for certain innovations may in fact be ten to fifteen years, the introduction of new technologies is a constant, ongoing process.

This is no idle speculation or a futile exercise in "what could have happened if?" In previous chapters the effects of technological innovations on the battlefield have been described. Just as a reminder, consider the effects of the longbow, the tank, the centimetric radar, and the atomic bomb on the side that did not have these technologies. It has also been shown that this kind of surprise could happen to everybody, including highly industrialized, technologically developed powers. Some of them were real surprises, based on previously unknown weapons, and some were cases of new, unexpected uses for a combination of familiar and well-known systems. If we sift out those surprises of the past that came about because of plain incompetence, do we have to really worry about something like this happening to us? (After all none of our generals or top scientists are incompetent.) It can be granted that technological battlefield surprises stem from innovations in science and technology. But by now we understand technology fairly well;

we read all the journals and follow the literature. So we should be current on all that is happening in the world of science, and with a little imagination, we should be able to predict fairly accurately what kind of weapons can realistically be developed and which weapons will remain science fiction for years to come.

In order to get a better handle on the dimensions of this problem—technological and scientific innovation, potential future weapon systems, and the effect of both on the shape and form of the future battlefield—try the following short mental exercise. Essentially this prediction, of the future battlefield, is a problem of extrapolation. Consider what is available today and make some educated predictions about the future. As every beginning engineer knows, an extrapolation can be somewhat improved if previous behavior of the function under analysis is taken into account. So take a moment to examine some previously developed technologies and their effect on the present and future battlefield. A famous or decisive war is a good pivotal point to anchor such an inquiry to, because it is easy to show what kind of completely new equipment was adopted during or after it. World War II, for example, could serve well, but it is too far in the past for our purpose. A second choice is the Vietnam War (1960–1975). A large number of new systems were developed and introduced during that war, but again, its beginning is too far in the past, and there is no convenient point during that war to serve as such a historical watershed.

Because of these considerations perhaps one should instead examine another war, the Six-Day War (June 1967) between Israel and several Arab states. It took place slightly more than thirty years ago, which provides a long enough, but not too long, time frame. More important, as its name implies, it was probably one of the shortest full-scale wars in history, and hostilities actually did cease after less than a full week, so no new weapons were developed during the war itself. Another important consideration for choosing the Six-Day War is the fact that both sides, the Israelis and the Arabs, were fairly well equipped with the latest weapons, which were produced in the west and east (respectively) at that time. So we know fairly well what was generally available then.

The Introduction of New Technologies

Let us examine then which technologies, and various types of equipment based on them, matured into general use within the time period referred to above; namely, from 1967 until recently.

Table 8-1: Technologies and their Applications

Basic Technology	*Military Applications*
Electronics	**Higher frequencies,** including millimetric wavelengths in all kinds of equipment, such as communications equipment and radars of proximity fuses.
	Advanced concepts, phased-array radars, synthetic aperture radars, improved beam control, advanced radar domes, conformal antennas, and configuration control to achieve stealth.
Electronic miniaturization	**Integrated circuits,** enabling smaller and lighter equipment, faster processing in computers, cheaper mass production, and decreased power consumption with its attendant savings in weight and volume.
Mechanical miniaturization	**Micromechanics and nanotechnology**, highly miniaturized components and systems, submicron powders for better and innovative materials.
	MEMS (Micro Electro Mechanical Systems), miniaturized electromechanical systems combining electronic systems, such as logic devices, and mechanical systems such as pumps, motors, and transmissions; some production techniques are based on electronics production practices.
	Advanced robotics and **advanced UVs** (Unmanned Vehicles).
Computers	**Personal and handheld computers,** for a host of applications, including C^3I (command, control, communications, and intelligence), fire control, logistics control, data fusion, decision making, maintenance scheduling

	and control, mission planning, and GPSs. Biological computers, utilizing living cells in order to enhance performance. **Artificial intelligence**, the use of computers for the fast analysis of problems and decision making, including analysis of past performance and mistakes. **Vehicle control,** including aircraft stability augmentation, and engine management and status control and information in both aircraft and land vehicles. **Virtual reality**, the creation of interactive environments for purposes of training and entertainment. **Autonomous ordnance**, capability to navigate and identify targets by specific characteristics, flight control of missiles, and unmanned air vehicles. **Digital communications**, enabling more reliable and encrypted communications. **Information warfare,** as an intelligence-gathering concept, as a means to deny intelligence to the adversary, and as a tool in infiltrating and sabotaging his computer-based systems, including the mass media.
Lasers	**Range finders** **Designators** **Laser-based gyros** **Lidars** (light radars or laser radars), as target and chemical warfare sensors. **High-energy lasers,** for use as weapons.
Electrooptics	**Thermal imaging** and **FLIR**, for surveillance, night-vision devices, and weapon guidance. **Adaptive optics**, controlled deformation of optical elements to overcome atmospheric distortion and achieve better laser beam control. **Optical fibers**, in communications, medical applications, laser gyros, and sensors.
Materials	**Advanced composite materials**, such as

	Kevlar, carbon fibers, metal fibers in various matrix materials (such as DU—depleted uranium—fibers) and ceramics matrices. **Advanced semiconductors,** for the electronics and electro-optics industries. **Advanced ceramics**, with better resistance to corrosion, mechanical stress, elevated temperatures, and shock, and cermetals, composed of ceramics and metals. **Stealth technology**, based partly on radiation-absorbing materials. **Advanced explosives and fuel technology** **Superconductivity** **Medical materials**, applications in joints, various implants, and artificial blood and skin.
Space technology	**Increased use of satellites,** for surveillance, mapping, communications, navigation (GPS), and weapons guidance, including microsatellites. The space shuttle and space stations.
Miscellaneous	Unmanned vehicles of all kinds, including unmanned combat air vehicles, nonlethal weapons, electromagnetic and electrothermal guns, high-energy particle and microwave beams used as weapons, improved storage batteries, and advanced stealth by shape control. And finally, theater ballistic missile defense and national missile defense.

Note: Obviously this is a partial list, enumerating only major groups of technologies and devices. A full list would have taken a book by itself, and in any case the purpose of this table is not to show all developments, but only to point out the variety and the breadth of the scientific disciplines involved and the current state of technology.[2] Furthermore, this table does not include combinations of technologies and systems and their potential military use. Take, for example, the various systems for aircraft defense against missiles. These systems comprise a variety of sensors, computers, and electro-optic devices integrated together to defeat the missile. Or in the same vein, highly integrated (radars, computers, propulsion systems, electronics, and electro-optics) systems to serve in the ABM (Antiballistic Missile) role. There are many more such combinations of technology. Moreover, military medicine has been barely mentioned. Further advances in this field will be based on sensors, computers, and biotechnology.

The Significance of the Above List

With the exception of space technology, which technically existed and was used before 1967, all the above are essentially newly introduced technologies that found wide-ranging uses. However, considering the above list as a whole, two facts stand out. In the last thirty years or so, on the average, a completely novel technology was introduced every twelve to fifteen months. This in turn gave birth to a new battlefield weapon or system (again, on the average) being introduced approximately every four to six months! Consequently, the potential for a totally new development, based on a new technology or on a combination of several existing ones, is significant and steadily growing. Admittedly, not all of these technologies are "war winners" in one single application. Furthermore, not all of them (such as materials or medicine) have direct war-fighting capabilities. But several of these technologies, when applied in the right place and against the "right" enemy, can have devastating effects. We have seen what the combination of stealth and highly accurate weapons did to the Iraqis in Desert Storm. An as yet unthought of combination can prove as efficient in a future war.

Since most of us grew up with all these systems around us, it is sometimes hard to understand their major impact on everyday life and on military operations. One may recall that the first handheld calculators appeared in the middle of 1973, and their price (in today's dollars) was equivalent to the price of a current advanced desktop computer system. Personal desktop computers started appearing around 1980 and had about sixteen kilobytes of memory. Before then, when somebody said "computer," he or she was referring to the mainframe version. To illustrate this point a little better, take a look at this issue from another angle. Imagine that tomorrow morning it is discovered that one of the above-described systems has simply disappeared. Let us take computers, for example.

It can be safely said that all military operations would grind to a screeching halt. Not only would most weapon systems become useless (except possibly the infantry's personal weapons), but all communications and practically all administrative functions would cease as well. This of course includes the phone systems and the power grid. The period of adjustment could take from days, for delivery of necessities (remember, today's cars rely upon computerized systems), to years, for redesign of most other weapon systems, but this time using slide rules.

This example is used only to show how much our technological environment has changed, how much today we take these technologies for granted, and how dependent we are on them. Another example is our present total dependence on satellites, for communications, observation (including weather), and entertainment. This growing dependence is never realized at the initial introduction of such technologies, but in a few years we find that we cannot do without them. In fact, the disappearance of all or most computing capabilities is a real threat and not just a theoretical exercise. It can be caused by EMP (Electromagnetic Pulse), which can be generated by the detonation of a nuclear warhead above the atmosphere. Depending on the altitude of the detonation, EMP can affect unprotected semiconductor devices, and computers, to ranges of thousands of miles.[3] Such effects can also be achieved by nonnuclear devices, the so-called EMP generators, which can be either placed near a desired target or carried by an aircraft or cruise missile.

Technical reports and discussion do not convey all the difficulties and ramifications of such an event. This is an excellent topic for a science fiction book that will illustrate the extreme degree of dependence, both in military operations and in our day-to-day life, on the above-listed novel technologies.[4]

WEAPON DEVELOPMENT TIMES

Weapon systems occasionally take a very long time to mature, but it can be shown that in many cases this is a problem of management practices and not of technological difficulties. It is true that in the end most of the viable systems or technologies are introduced, but usually after a lot of time has been wasted, and if a war broke out, these systems would not have been available. Apart from the delays in the timely introduction of (presumably) a desirable system into the hands of the troops and the problems this can cause,[5] these long times have other, occasionally less tangible effects, particularly in peacetime. For one, it has been shown that as the time involved in developing a technological project increases, two things happen. The first is that the price of development goes up. This in turn is viewed askance by the regulating or financing agencies, and consequently, the chances of the project being cancelled also increase.[6] Furthermore, the late introduction of a weapon system has a wide variety of undesirable effects on preparedness, training, doctrine development (in the case of a novel system), and timetables for other weapon systems. Long development times and delays are not a law of nature, and like any technological failure discussed previously are essentially man-made. To prove the point, it is worthwhile to look at some previous technological development projects and their timetables.

In 1915, the British Admiralty issued a requirement for a new airship to be used in submarine hunting. It was to be a two-man craft carrying wireless equipment and bombs and was to have a velocity of fifty miles per hour and eight hours' endurance. The first prototype was finished by the Royal Navy in three weeks, and manufacturing then passed to a civilian contractor.[7]

From an idea at the end of 1934, to a first test in February 1935, the British Chain Home radar matured to a battle-tested and successful system by the end of 1940.

In July 1940 a new design for a tank landing craft (LCT), to carry five tanks, was prepared in three days. The first of these were produced in three months.

In 1940 the RAF ordered the P-51 prototype from North American Aviation, stipulating that it was to be delivered in 120 days. It was delivered in 117 days. Admittedly it became a "real" fighter only after the RAF installed a better engine, but presumably if the Merlin engine had been specified from the start, North American could probably still have done it in time.

In June of 1943, Lockheed was contracted to build the P-80. The contract called for 180 days development. It took Kelly Johnson, the project manager, 143 days to roll out the first prototype. The "Skunk Works," under Johnson's leadership, continued this tradition. The U-2 was developed in eight months and the SR-71 in four and a half years. Development of both these aircraft involved major technological challenges. The U-2 was to fly higher than any air-breathing engine ever flew before, and the SR-71 design had to overcome severe heating that stemmed from its sustained supersonic speed.

Finally, the Polaris missile (to be fired from a submerged submarine) was developed in less than four years, and the Moon-landing project was accomplished in slightly more than eight years, and of course the technological breakthroughs involved in the last one added a significant page to world history.

There are of course numerous other examples, but these will suffice to illustrate the point. Properly managed, a development program does not necessarily run for decades, and the complexity of the system has nothing to do with development times.

So two conclusions can be reached. First, a new technology or new device based on a new technology may pop up, without warning, in any country with a reasonable technological and industrial base. Such a device may have great battlefield potential and in effect may constitute a battlefield surprise. Second, our notions about what is possible and what is not must become much more flexible, and previous experience is not necessarily of much help. Since futuristic ideas are often defined as science fiction, a brief look at this topic follows.

IS IT REALLY SCIENCE FICTION?

Introduction

Vannevar Bush (see chapter 6) considered as fantasy many novel ideas that we now take for granted. When dealing with future developments, quite often one hears the definition "science fiction," usually as a derogatory definition of a concept whose time has not yet arrived. This brings up the somewhat knotty problem of the borderline between hard science and "science fiction," not in the literature, of course, but in the day-to-day judgment of novel ideas. Since the distance between science fiction and future technology is obviously very short, perhaps these terms can be clarified a little. The issue is not just an idle philosophical question. In any research and development organization that wants to stay at the cutting edge of technology (no matter whether it deals with future concepts for weapons or with communications and entertainment technology for the mass media) this would be a crucial question. When a very advanced concept is considered for research or development, the answer to this question will actually determine the organization's willingness to get into it. So first we have to define what is "science fiction" and second, when does a new idea stop being science fiction and become open for dicussion, maybe even sent for publication to a reputable journal, with the hope that the editor will actually consider it?

What Is Science Fiction?

Let us start with the first question: what is technological science fiction? Is it only something in the domain of mad scientists with tousled hair and wild eyes, like the scientist in the *Back to the Future* movie, or is it any piece of equipment that cannot yet be ordered from a catalog or that Sony did not get around to developing? Is science fiction defined by time; in other words, is there anything that today is considered science fiction that will be on the shelves twenty years into the future? Finally, does science fiction deal with concepts or only with hardware—maybe some futuristic piece of equipment to do something quite mundane?

A look back at some technological advances that became realities in very recent times will help to answer some of these questions, put our thinking in some order, and possibly define some better guidelines for the future. The following examples were chosen from dozens that are available. The choices, however, were dictated by the public's general familiarity with the topics and by the presented technologies' effects on our lives.

A Short History of Space Flight

On October 3, 1957, most of the educated public was united in considering flight to the Moon a fairly wild brand of science fiction, and the only question in the public's mind was about the degree of sanity of those who believed in this fantasy. Yet on the next day, October 4, the Soviet Union launched its first Sputnik and caused some of the most profound scientific and political upheavals in history. Apart from the sudden emergence of the Soviet Union as a first-class world power with a proven expertise in one of the most—if not *the* most—technologically complex fields of engineering, this feat had broadcast another powerful message: the United States was no longer secure behind the "pond"—the Atlantic Ocean. The Soviets, in fact, said words to that effect earlier that same year, but at the time the pundits argued whether this was for real or just a Soviet hoax. Since a satellite launcher also has the capability to carry a reasonably sized warhead to almost anywhere on the face of the earth, this question at least was now answered. The Soviet achievement thus pulled the rug out from under one of the mainstays of the American defense doctrine, which assumed that any Soviet attack on the continental United States would have to be carried out by bombers or by submarines that would have to get fairly close to the shore. But since there were no suitable long-range Soviet fighters (to escort the bombers) in existence at the time, it was believed that the Soviet bomber force (sans fighters) was not a terrible menace and with the aid of an appropriate radar warning net could be easily handled in case of attack. After all, hadn't the United States already been through a similar experience over Germany? As for the submarines, they would have to surface to fire their weapons, and it was assumed that the navy could deal with that problem. But a ballistic missile force was altogether a horse of another color. There was no way (then) to stop a ballistic missile in flight, and the existing radar systems could not even detect such missiles.

Psychologically, the American public was faced with another issue. The simple fact was that this Soviet success came on the heels of several failed American attempts to do the very same thing. The Soviet success was more startling in that while the American failures were quite publicized—some of them in real time, so to speak—the Soviets' achievement came as a total surprise, with not even a hint of their intentions or preparations leaking out. This of course raised a host of other questions. First, if the Soviet Union managed to hide the fact of its preparations for a satellite launch, there were two possibilities. Either the Soviets managed to hide all of their test shots and (presumably) their failures from the whole intelligence-gathering apparatus of the United States, or they succeeded on their very first attempt, and nobody wanted even to contemplate which of the two was worse. Second, what were the Soviet capabilities in the missile production field? One thing was sure, though, particularly after the flights of Leica (the dog) and the cosmonaut Yuri Gagarin: space flight was here to stay, and with it some of the worst nightmares about intercontinental nuclear warfare shifted from science fiction into official defense planning. This was not yet a flight to the Moon, but it was obvious that such a trip was a reasonable goal for any nascent space-faring power.

To America's credit the country recuperated remarkably well. On January 31, 1958, less than four months after the first Sputnik, a United States Army team, headed by Wernher von Braun, successfully launched the first "Explorer" and the United States joined the "Space Race." On May 25, 1961, President Kennedy formally announced the United States's intention to send a man to the Moon and bring him back, and to do so within the decade.[8] The reader is reminded that on July 20, 1969, the *Apollo 11* crew did land on the Moon, came back safely, and opened the way to lunar exploration.

It is worth noting that von Braun, head of the team that launched Explorer, was a rocket enthusiast from his early adulthood in Germany. In 1932 (before the Nazis came to power), at the age of twenty, he joined the German army's weapons development branch in order to work on long-range rocket technology and subsequently became the head of the V-2 project. At the war's close, he and most of his team escaped to the American sector in Germany and eventually were brought to the United States as part of "Operation Paperclip." This was an effort (very successful as it turned out) to learn as quickly as possible from these prisoners all they knew about the advanced technologies developed in Germany during the war. The Soviets ran a similar project that helped them enormously in these fields. An argument was advanced that von Braun used both the German army and NASA to further his own dreams of space travel. In fact, in 1944 he got into a spot of trouble with the Gestapo in Germany, who suspected him of this very thing. Since in this case he was of immense value, both to the Germans and to the Americans, it might be said that both fields of science benefited.

Space Flight and Science Fiction

The purpose of the above narrative is to underscore three points. First, for most of the world's public, space flight moved literally overnight from the twilight zone of science fiction into the realm of hard science. Presumably the engineers, in both the Soviet Union and the United States, who were working on these rocket projects, were long-standing believers, but one can still wonder about their political bosses.

Second, within three years from the first Soviet Sputnik, space flight in the United States advanced from mere attempts at satellites and Earth orbiting to the status of a national project aimed at the Moon and endorsed by no less than the president himself. As an aside, it might be interesting to note that Kennedy was more interested in beating the Russians than in space exploration.[9]

Finally, today our lives (in many areas literally) are dependent on space flight. Weather information, communications, navigation of ships, aircraft, and even GPS for hikers, entertainment, and the myriad uses of satellites for military purposes are inconceivable today without the various satellite systems now in use. Space flight changed from science fiction into a multibillion-dollar industry in a very short time.

Some of science fiction has been enmeshed in real science in curious ways. Arthur C. Clarke, the noted science fiction writer, published at the end of the forties a story involving geostationary satellites. These are satellites orbiting at a height of about twenty-one thousand miles. At this height their period of rotation is twenty-four hours, and if they are in an equatorial orbit, they are in essence continually over the same spot above the earth's surface. This is a feature of many communications and surveillance satellites. When the first "Geo" satellites were proposed for real use, Clarke claimed that he deserved royalties for being the first to advance this concept. But because more than seventeen years had passed since the publication of Clarke's story, his claim was deemed void.

A Short History of "Death Rays"

The first instance of the use of the concept of "death rays" is attributed to Archimedes, the famous mathematician who lived in Syracuse, in what today is Sicily. In the second Punic War (214–213 B.C.), the town sided with the Carthaginians and came under siege by a large Roman force and a Roman fleet. Archimedes constructed various machines of war that played an important part in the defense of the town. It is also reported that Archimedes constructed large mirrors and used them to put fire to the Roman fleet. Although the town was eventually conquered and Archimedes was killed, the story lived on, and in 1973 a Greek engineer organized a test to check the veracity of the story. He took seventy Greek sailors, equipped them with brightly polished bronze shields, and illuminated with them a small wooden boat at a range of about two hundred feet. In two minutes the boat was on fire. Presumably this is what happened to the surprised Romans, but we can also assume that after a while they learned to be careful and stayed out of range. Unfortunately, we do not know how much this invention actually contributed to the defense of the city.

Through the ages all sorts of "magicians" were credited with the power to create damage without any physical agency—by remote control, so to speak. In 1898 "Martians" joined this mystifying group. In H. G. Wells's fictional account of an invasion of Earth from Mars (*The War of the Worlds*), the invaders were equipped with such tools. Their weapons consisted of a complicated-looking device that produced a heat beam, which vaporized everything it touched. In fact, the description of the beam is very much reminiscent of a current high-power laser beam. Ultimately, Earth was saved because the invaders succumbed to earth bacteria—a possible prediction of biological warfare. After the publication of this book, the floodgates opened and the fictional literature, and later the movies, were regularly describing all sorts of death beams based on all kinds of weird physical principles. All of this was of course fiction, but everybody agreed that if something like this could actually be developed, it would be a superb weapon. Possibly under the influence of this literature the Air Ministry, in 1935, asked Watson-Watt, the British radar pioneer (who in turn asked Arnold Wilkins, his assistant), to investigate the real-life possibility of a death ray (see chapter 3). Broadcast radio beams, after all, had all the requisite qualifications to be death rays. They did broadcast power, they could be aimed, and they were based on solid science. Unfortunately, they also were too weak to cause real damage, because the technology was yet immature.

The advent (from 1983) of the "Star Wars Project" (the SDI—Strategic Defense Initiative), with its huge budgets, enabled the research into death rays to be resumed seriously, this time aimed specifically at hitting and downing ballistic missiles. In the time between the days of Watson-Watt and those of President Ronald Reagan, technology had advanced and now there were several possible technologies for the "death ray" weapon. There were free electron beams, neutron beams, microwave beams, and last but not least, laser beams.[10] Because of its wide utility in other fields, laser technology was the most advanced and thus got the most attention of the researchers. One advantage of the laser as a weapon was that even at relatively low powers it could damage a human eye, and this was the first intended use for it, when thinking turned toward antipersonnel devices. After the Falklands War, a sharp-eyed photographer caught some strange-looking boxes mounted on the mast of a British destroyer at port. Somebody probably forgot to mount the covers on those boxes. The pictures were published, and it turned out that these boxes were laser devices intended to dazzle Argentinean pilots diving on the ships.

But this was not yet a true "death ray." Further advances in the relevant science and the acute need to shoot down fast-moving targets finally bore fruit, and in several tests artillery rockets (Katyushas) were recently shot down in flight by an experimental laser weapon. Admittedly, the weapon was ground based, so the whole system was less weight sensitive, but on the other hand this rocket has a fairly thick steel casing, and to heat it sufficiently to cause an explosion required both considerable power and considerable accuracy. Finally, the United States is now engaged in developing the ABL (AirBorne Laser), which is to be flown on a specially modified Boeing-747 and intended to shoot down ballistic missiles at ranges of hundreds of kilometers.

A wide variety of other smaller, handheld and vehicle-mounted lasers are being developed, and the only obstacle to a wide distribution and use of these kinds of weapons is the problem of power supply for the laser. But the high-energy-beam weapons are here, and in essence the fictional death ray—moving at the speed of light, and having the ability to vaporize a target—is here. Admittedly, it took somewhat longer than space flight to be fully realized, but this is not really the issue.

Biotechnology and Science Fiction

The third area of great scientific advances, which until a couple of decades ago was featured only in science fiction literature, is that of biology and the biological sciences. Specifically we are referring to the area of genetic engineering and cloning. While at first glance the relevance of these fields (compared with the previous two) to military technology seems a bit tenuous, two points must be made. First, this is an attempt to make an observation about the relationship between science fiction and science fact. Second, biology does have a great relevance to the military art, and not in the generally accepted form, but that will be addressed in a moment.

Without going into the ethical questions, it seems that cloning is by now a fairly proven technology and, one way or another, is here to stay and develop. Cloning is of course an old topic in the science fiction genre and thus qualifies to be included in this discussion. Genetic engineering deals with the ability to change original characteristics of a living cell, and by projection, of a whole organism, by tacking on certain characteristics from other cells or organisms. In the past these techniques always dealt with the appearance of endless (cloned) hordes of warriors with the attributes of supermen, at the service of some evil ruler or a mad scientist. However, such an application for the science of cloning is really most inefficient, actually wasteful. Without going into wide-ranging predictions, it should be pointed out that this kind of technique, when applied to several types of animals, will reap considerably more military dividends than applying them to humans. Furthermore, genetic engineering can have important applications in the field of treating combat injuries, especially in the repair or replacement of badly injured organs.

A Practical Guide for the Future

So what is "science fiction," and when does it stop being science *fiction* and become science *fact*? Is it a question of theory, of actual gadgetry, or does it only depend on attitudes?

To simplify things a new definition for science fiction is proposed that in the future, when discussing a battlefield innovation, may put things in the right perspective: "Science fiction is any device or procedure, the technology of which has not yet been proven in the laboratory." The reader will notice that this definition has nothing to do with size, form, power consumption, efficiency, reliability, or price. If the principle (technology, if you will) was proved in a bench test and is repeatable, it is science, whether original or transformed from science fiction. It may yet have physical limitations of weight or range, just to pick two of the most notorious in many instances, but nevertheless the principle was proved. From there on it is only engineering. To emphasize this last point, here is another example. Until the last decade of the nineteenth century, heavier-than-air flight was considered by many to be science fiction, and many a well-reasoned and mathematically rigorous paper was written to prove it so. It turned out, however, that the whole problem revolved around the lack of a suitable engine. Once that was solved and the Wright brothers invented the ailerons, or at least their precursor, wing warping, the road to the jumbo jet and the space shuttle was open.

And here the circle closes. What today is a wild notion, based on science fiction, may suddenly mature into a useful technology with undreamed of capabilities. Throughout history, almost any innovative technology had military implications, although as we have seen, not always immediately apparent or acceptable. But a resolute application of a novel technology may lead to a surprise, be it scientific, commercial, or military. Such "technological surprises" are the subject of the next chapter.

9 Technological Surprise and Technological Failure

In war, there is no second prize for the runner-up.

General Omar Bradley

BATTLEFIELD SURPRISE

Classic Military Surprise

Tactical or even strategic surprise is as old as warfare itself. Several examples are already discussed in the Bible, like Gideon and the Midyanites (Judges 7:19–25) and Jonathan and his squire at Mikhmash (I Samuel 14:13–16), both involving night attacks by numerically inferior forces. Every green squad leader has heard of it, and it is one of the "principles of war" discussed both by Clausewitz and Sun Tzu.

Surprise on the battlefield has the most unwelcome, if not devastating, consequences for the surprised one. Surprise can arise from carelessness (Pearl Harbor), preconceived ideas (Israel in the Yom Kippur War), or complacency (the Allies in December 1944 and the Battle of the Bulge). It can be achieved by careful planning and execution (the location of the D-Day landings) or by intentional deceit (the Sicily invasion and the "man who never was").[1] It can also result from sheer luck combined with audacity such as the capture of the Remagen Bridge on the Rhine on March 7, 1945, although in this instance the luck was supplied by German carelessness. Surprise can be planned for by a bold move (Operation "Market Garden" in September 1944), but turn into a disaster because of poor planning and politics.[2] Finally, surprise can be achieved by a novel approach route that is considered impassable, like the German attack through the Ardennes in May 1940 and again in December 1944, or the Israeli Defense Force flanking the Egyptian army in 1949, by moving through supposedly impassable dunes, but actually traveling over an ancient Roman road.

Technological Surprise

These were all cases of military surprise that almost always depended on an unexpected action or movement of troops, and as the quotations from the Bible demonstrate, this concept of surprise was known to warriors from early history and probably before. But there is another type of military surprise, when unexpected "artifacts," usually weapons, appeared suddenly on the battlefield, quite often rendering helpless the other side, which lacked these new artifacts. This is what may be defined as a "technological surprise." But the concept of "technological surprise" was seldom treated in the literature, even when it was obviously the main cause of the outcome.

Not that the effects of radically new weapons were not talked about and elaborated on at length in the literature. What was discussed, however, were the long-range (timewise) effects of novel technologies (technological innovations, if you will) on the battlefield and on the development of the art and science of warfare. But only rarely (and only in the most recent history) is there any description or analysis of the actions, and the reactions, of the combatants at the time of the appearance of a superior or novel weapon, whether in a battle, a campaign, or even a war. There are a lot of discussions, for instance, of the effects of powder artillery on castles, on feudalism, and on the social fabric of the time. Likewise, after 1945, nuclear warfare was widely discussed. There is, however, no description of the thoughts and actions of the gunners and their master, on the one hand, and the castle's lord, on the other hand, on that first evening after the guns fell silent. Nor do we know much of the thoughts and actions of the castle lord's colleagues when that particular siege ended with the rather predictable results.

This particular event is lost in the mists of history. We do not know which castle was the proving ground for the new gun technology or when this particular battle occurred. However, considering the sheer number of different weapons that were developed in history, the occurrence of a technological surprise was not so rare an event as we who have been born into the age of technology first tend to assume. The number of different implements of war invented in the last several thousands of years, from the first clubs and javelins to the present day, and including every weapon in between, reaches probably several tens of thousands. Even if we consider only original concepts and not variations and improvements to an existing system, the number will still be very large. This translates to the fact that on the average, a totally new weapon made an appearance every several years and resulted in battlefield surprise.

The above reference to "artifacts" is intended to highlight the fact that a technological change on the battlefield can be achieved by means other than obvious "weapons." For instance, the first use of stirrups by the Avars (a tribe of nomads) of central Asia sometime in the sixth century. This rather uninspiring device—probably invented to make life easier for the rider by providing him with a simple, single-runged "ladder" to climb onto the horse—turned out to be a most important tool and was copied by everybody who saw it. It converted cavalry from just a fast-moving infantry ("dragoons" in later terminology) into a shock force, capable of actually fighting from the saddle (in itself an important invention), and set the stage for cavalry warfare for more than a thousand years. In more recent times, the same thing happened to other nonmilitary technologies that were "drafted," so to speak, into military service: the balloon, the telegraph, the internal combustion engine, the radio, and the airplane. Both the railroad and the cargo airplane (although not weapons by themselves) were used numerous times to achieve military surprise by rapid movement of troops to unexpected locations. The most recent example of such "militarized" devices is of course the computer.

TOTAL SURPRISE AND SELF-INFLICTED SURPRISE

Technological surprise can thus be defined as the appearance of new equipment that contributes to the fighting efficiency of the force who employs it, and that is employed against an opponent who was not prepared for this event or its consequences. The history of warfare is replete with examples of military forces who fought actions, or even went to war, without knowing that the enemy was equipped, or was becoming equipped, with technological devices against which the force's existing equipment did not provide an answer. This may be defined as a "total surprise." Battlefield systems were developed by one side in strict secrecy, and the secret was kept until said equipment was introduced in action. For example, the acoustic torpedo, developed by the Allies to fight the U-boats, the upward-firing guns mounted on the Me-110, and the Dam Busting bomb, developed to breach the Ruhr dams. Other examples include the use of window and the jamming of the (Soviet-made) Styx sea-to-sea missiles by the Israeli navy during the 1973 war.[3] The ultimate such weapon was of course the atomic bomb. The awakening to the facts of the real world was usually very rude. While such a development can be called an intelligence failure, it should be remembered that even the best intelligence services are not omnipotent and, depending on circumstances, can come up short. But as has been shown, in many instances there was enough circumstantial evidence to warrant a more serious investigation, and the failure was not in neglecting to carry out such an investigation but incorrectly assessing the already available data.

A second form of surprise results from one of the forms of technological failure previously discussed. It involves technological developments or weapon acquisitions by one side, of which the other side—or for that matter anybody who bothered to look around, read the professional literature, or visited a technical exhibition—was well aware. Occasionally, not only the existence of the new weapon was known, but a lot of useful information concerning it was available as well. Often, however, the capabilities of this new weapon and potential battlefield scenarios where such a weapon could be used were not considered together. There were also cases when one side actually encountered the said weapon, or even used a similar version of it in the past, and later ran into it and still was surprised by the results. This may be defined as a "self-inflicted surprise." The longbow at Crecy, the radar at the Battle of Britain, and the Israeli Defense Force's encounter with the Saggers are examples of this kind of self-inflicted surprise.

THE LIMITATIONS OF INTELLIGENCE GATHERING

As shown in the previous chapters, technological surprise rarely stems from an intelligence failure. Most of the examples cited dealt with bad understanding or misinterpretation of known facts (sometimes facts of public knowledge), conservatism, and misguided economics. The example of the Austrians (see chapter 1) who did not want to order a modern rifle because they had just finished retooling their factory to produce another, albeit obsolete, one is cogent in this regard.

Can more efficient intelligence gathering diminish the danger of a total technological surprise? The answer is not a definite yes or no. Intelligence is important because it can shed some light on some of the threats and possibly avert a total surprise, but no intelligence service can be considered reliable enough to be able to provide the goods in time. And what does "in time" mean in technological warfare scenarios? Weeks, months, or years? Furthermore, the warning itself is not sufficient when it comes to a technological threat. As has been demonstrated in numerous examples, the problem usually is not with data gathering or even interpretation of the data, but in managing such data to the best effect and doing something useful with it. Two more examples will serve to illustrate this point. Chapter 4 (note 29) noted the capture by Rommel's forces of a Wellington airplane equipped with an antisubmarine radar. Rommel did not bother to inform his superiors in Berlin of this find, either because he wanted to use this airplane himself, which he did, or because he thought that the find was of no importance—Berlin probably had one anyway. Berlin finally learned about the airplane and immediately demanded it. Based on the captured radar, the Metox receiver for submarines was developed. A similar example, this time on the American side is cited by Samuel Morison.[4] The Japanese "Long Lance" torpedo was better than anything the United States possessed, even when American torpedoes worked as designed. It had a more advanced motor, a warhead twice as heavy, and more than twice the range—in fact, greater than a battleship's guns. It caused much grief to the U.S. Navy and was instrumental in several Japanese naval victories. Nobody in the U.S. Navy in the Pacific knew of this torpedo or its performance. That is understandable. But then one of these torpedoes apparently washed ashore on Savu Island (in the Guadalcanal group) in January or February 1943. It was locally taken apart and the findings sent to Fleet Intelligence, and there the data disappeared and nothing but rumors filtered back. Even Admiral William F. Halsey's staff did not know about it. The rumor was brought to the attention of an admiral who was planning a battle (Kula Gulf, July 1943), and he was warned of the torpedo's excessive range, but he dismissed it as "scuttlebutt." In the coming battle, the U.S. Navy lost a cruiser and two destroyers, the cruiser definitely being lost to this torpedo.

This story has a sequel, however. In 1996 an article was published about the Office of Naval Intelligence (ONI) of the U.S. Navy in which the author claims that in 1940 the navy in fact knew all there was to know about the "Long Lance." However, when the data (its "phenomenal" speed, range, and weight of warhead, compared to western weapons) was submitted to the Bureau of Ordnance, they dismissed it as "impossible." Part of the reason for this attitude was the fact that the Japanese torpedo relied on internally stored oxygen and not on stored air. Since both the U.S. and Britain had not yet mastered this oxygen technology, the "experts" concluded that neither could have the Japanese.

In this case, too, the intelligence-gathering agencies did their job—and under the circumstances better than might have been expected—but to no avail. Readers may notice the similarity of the attitude of the Bureau of Ordnance people to that of the German scientists who dismissed centimetric radar as science fiction.[5]

THE EFFECTS OF TECHNOLOGICAL SURPRISE

In the short run a technological surprise can be treated like any battlefield surprise. The particular operation can be abandoned or an effort will be made to overcome the opposition, either by a prodigious expense of resources, by some sort of a juggling act against other demands, or by using other means to even the odds. The surprise and plight of the American tankers when faced with the more effective German eighty-eight-millimeter tank guns in Europe is a good illustration. The United States Army had a great numerical advantage over the German army, and although suffering heavily the U.S. troops pushed the German panzers back.[6] They were aided to a large extent by fighter aircraft that provided close air support. This in turn was possible because by that time the Luftwaffe had been completely swept from the skies over Normandy. On D-Day the Luftwaffe had only eighty operational planes to oppose the invasion.[7] Two months later not one single German airplane made an appearance in the crucial armor battles of August 1944.[8]

In the long run the situation is different. As pointed out above, technological surprises are caused by technological failures. The best course of action is to adopt a suitable technological counter, but this doesn't always work. If a radar system is successfully jammed, more radars of the same type will not solve the problem. A new radar system or a major modification will be needed, and these cannot be pulled out of a sleeve unless one is very agile in his thinking, decision making, and organization.

If this is impossible, if resources to be sacrificed are not unlimited (or this will not help because the technical disparity is too wide) and there is no other way to circumvent the problem, the result is defeat, either local or total, forcing a concession in that particular battle.[9] The Germans who desisted from daylight attacks against Fighter Command bases (and consequently gave up Sea Lion—the invasion of England) and the slowdown of the American daylight bombing in the fall of 1943 are two examples of such defeats. A doctrinal innovation may sometimes serve as an interim solution to a technological surprise, or for that matter to a technological inferiority, but its effect will often be short-lived, costly, or both. This is because new tactics can be countered by organization and training, but the original cause—the technological innovation or superiority—still lurks in the background. Several examples will illustrate the point.

General Claire Chennault, of "Flying Tigers" fame, taught his pilots to use the superior speed of their P-40s to dive through Japanese bomber formations and not to tangle with the more agile Zero fighters. Since Chennault's purpose was to stop bombing attacks, this worked admirably. On the other hand, consider the following. In 1948, during the British evacuation of Palestine, several key fortifications (the Taggart police stations) were turned over to the Arabs. The Israeli Defense Force, not having artillery, sent explosives-carrying soldiers at night to approach these fortifications and blow the walls down. The strategy sometimes worked, but more often it did not. The machine gun and barbed wire combination of World War I was a technological impediment to movement. The Germans tried a technological solution, gas, but failed because the attempt was not managed seriously. The Allies tried the tank, but at first failed because of poor understanding of its qualities. The Germans then tried a tactical solution, the "Hutier Technique," which involved infiltration by small units and no preceding artillery barrage. Initially, these infiltration tactics achieved spectacular successes, probably because of their novelty, but once the Allies learned to cope with them they petered out.[10] The only solution that finally worked, and which shattered the concept of a static line, was the armored force. Today this armored force is coupled with airborne troops and helicopter gunships, again an essentially technological innovation. Innovative tactics, however, will hardly stop an armed helicopter; a better weapon specifically designed for the purpose might.

Can a military recover from technological surprise? This of course depends on circumstances. There is not much that can be done after a nuclear strike. Nor can be much done if the cause of the "surprise" stays hidden. In this case combatants are lost, and the causes might not be discovered for a long time. This was what happened when the Allies introduced centimetric radar and later the antisubmarine acoustic torpedo. The conservation of the "secret" can be aided by a skeptical opponent, such as the intelligence experts in Bomber Command who refused to credit reports about upward-firing guns, or the French intelligence interrogators who discredited the information about an impending gas attack. Finally, there are the surprises that little can be done about, like the appearance of the single-engine escort fighters accompanying deep-penetration raids into Germany. Goering initially discredited such reports, because at that time the Germans still believed that such a fighter was technically impossible.

Can Technological Surprise Be Exercised Like Any Military Operation?

Training and exercises are excellent preparations for future eventualities. Military commanders are generally trained to react to battlefield surprises, and most of such operational surprises can be simulated in training. Except for equipment breakdown, which also is practiced in many branches, technological failure (in the presently suggested meaning) is never practiced, and it is doubtful if it could really be simulated or practiced. In order to do so, a notional futuristic weapon (or "situation") would have to be introduced into the exercise, and one can imagine the general reaction to such a suggestion.

SOME CONCLUDING REMARKS ABOUT TECHNOLOGICAL FAILURE

The Effects of Technological Failure

The technological failures that have been described (or briefly mentioned) in the previous chapters can thus be divided into three approximate groups:

A. Those failures that actually caused or might have caused certain total defeat.
   1. The German refusal to acknowledge the nitrates problem prior to World War I.
   2. British inability to effectively counter the submarine threat in World War I and a similar failure of the Allies to anticipate it during World War II. (The Allies were probably saved by several glaring German mistakes.)
   3. The German lack of appreciation for the role of science in a modern war. This in turn led to superficial conclusions about Allied capabilities in many fields.
   4. The blind faith, or overconfidence, of the Germans and Japanese in the unbreakability of their ciphers and codes.
   5. Iraqi disregard of advanced technological capabilities of the United States forces in 1991.
   6. German lack of appreciation for the role of British radar in the Battle of Britain.

B. Those failures that voided the possibility of an easy victory, or at least a reasonably certain one, or materially changed the strategic situation.
   1. The initial (1915) German failure to appreciate the potential effect of gas warfare.
   2. The slow British acceptance of the tank, and then its hasty introduction.
   3. The Germans not introducing drop tanks for single-engine fighters, and their trust in the performance of twin-engine fighters.
   4. The German delay in the introduction of the snorkel and the jet fighter.

C. Those failures that caused unnecessary casualties and wasted resources and time.
   1. The United States Navy's failure to quickly rectify the torpedo problem.
   2. The United States Army's failure to adopt the seventeen-pound tank gun. This is a marginal case. Under different circumstances, such a failure could have led to a defeat.

3. The United States Army Air Force's delay in the introduction of drop tanks, and the development of the long-range escort fighter.
4. The Allied delay in the introduction of window.
5. Bomber Command's disregard of reports from its own operations research people.

Obviously, it is difficult to unequivocally assign a particular form of technological failure to any one of these groups. Moreover, many such events were later mitigated or amplified by other causes, further clouding the issues. Certain actions from World War II (and even earlier periods) are still debated today by professional soldiers and historians.

Some Characteristics of Technological Failures

A technological failure that manifests itself during actual warfare will almost always result in an operational failure, though the operational failure might not always be immediately traceable to the technological failure, and depending on circumstances, it might never be so traced. Rommel, who several times was outmaneuvered by the British forces during the fighting in the Western Desert, became uneasy about communications security. He should have been, since all the radio traffic between his headquarters and Berlin, enciphered by the Enigma, was being read by the Allies, almost in real time. Usually the Allies acted on such information only when there was corroborating evidence from another source. Once, however, the Allies had to take a chance and attack an Axis naval supply convoy moving in a fog that could not be discovered by an "accidental" reconnaissance. Rommel complained about this to his superior, Field Marshal Albrecht von Kesselring, who conveyed these suspicions to Berlin, but after an "investigation" was told that this was simply impossible, and so the leakage continued.[11]

Operational planning may take from hours to weeks, depending on the scope of the operation, but once the action has started, poor planning or execution show up very quickly. On the lower, tactical level, failure will come almost immediately on the heels of a bad decision. In most wars, a local tactical defeat may be contained and, except for its effect on casualties and on timetables, may be weathered. While such a failure is unpleasant, a quick and flexible response can occasionally snatch victory from the jaws of defeat. Contrary to this situation, a technological failure that takes years to "brew" will manifest itself in the most unexpected way and in most cases is impossible to efficiently rectify on the spot.

Not all the technological failures were, or will be, total disasters. Nor were all of them irreversible events that once entered into could not be rectified. The failure of both the Germans and the Americans to adopt drop tanks for their fighters when they were most needed is a case in point. With a little more diligence, both ongoing failures could have been terminated almost instantly, as was eventually the case with the Americans. But under the right circumstances even the smallest technological failure might cause a chain of events that very quickly will grow into the proverbial snowball, with unforeseen consequences. Nowadays, in the western democracies such an event may have far-reaching consequences. Minute military setbacks (that in the past would have been barely noticed) can erupt into political upheavals that would change national policy. Furthermore, during a war it is hard (or there simply is no time) to sit and analyze such a development in a quiet and orderly manner in order to find suitable countermeasures, technical or tactical, particularly if the war is a short one, as seems to be the present trend. It is doubtful that there will be a repeat of the recent eight-year-long (1980–1988) Iraq-Iran war.

In many cases the side that was affected by a surprise weapon did not realize what was wrong or even that something *was* wrong, even though the failure had a severe impact on his capability. The German misunderstanding of the role of radar in the Battle of Britain is an example. And although it was not a case of a "surprise," Nazi Germany's attitude toward its scientists is an example of not realizing the gravity of the situation. In other cases, the effect of the technological failure does not lead to a total and immediate operational failure, but may sap resources in an insidious way with cumulative effects on actual war fighting or morale. Eventually there is a slow awakening and realization that all is not working as expected. The failure of élan in face of the machine guns, the mounting losses during the daylight bombing raids (and the losses of Bomber Command in its night raids), the constant losses of the German submarine arm in the Battle of the Atlantic, and the losses of the American tank crews in France all led to a gradual crumbling of confidence in the superiority of one's own forces.

Potential Political Considerations with Regard to Technological Failure

Development times for new military equipment (including its production and absorption by the services) are extremely long. A lead time of a few years during peacetime may become almost impossible to close after commencement of hostilities, and this may give the adversary a window of opportunity that he might wish to exploit, either by forceful diplomacy or by actually initiating a conflict. While it was not a clear-cut case of technological failure but rather more of a political failure, this is what happened to the British and the French in the thirties. In 1935, Britain, which wanted to use its Royal Navy to stop the Italian move against Abyssinia (today's Ethiopia), was intimidated by the Italian air force and backed off, but with a flash of foresight (watching Germany rearm) the British in turn started a vast rearmament program of its own. In the meantime, the Germans used their lead time in rearmament well. The Luftwaffe so frightened the leaders of Europe that time and again they chose to acquiesce to the German demands. The British rearmament program came almost too late; most of Europe fell to the Nazis, and Britain was saved from conquest in a closely run contest that climaxed in the miracle of Dunkirk and the Battle of Britain. Both of these ended as they did largely due to mistakes on the part of the German leadership. A present example is the growing proliferation of ballistic missiles, with only a start of deployment of antiballistic missile systems and the ongoing debate in the United States about National Missile Defense. Consequently, in the meantime those who have the ballistic missiles might feel that an existing window of opportunity might close. Under such circumstances, there is a strong temptation to "use it or lose it."

After the hurdle of the transformation of a novel technology into a practical system has been passed, the questions of utility and price still remain. In many instances, particularly since the end of the Second World War, the scientific proof of a new concept comes from the civilian market, and it is occasionally difficult to convince the politicians and the military authorities of the importance of novel, potentially doctrine-changing technologies. Here of course enters also the question of price. The trouble is that the price of a novel technology or weapon system in itself has no meaning. It has to be weighed against utility, and here is the rub. How do you assign a financial value for the utility of a future weapon on a future battlefield and against an as yet unknown enemy? In most cases even operations research cannot help here. Operations research can do wonders after a war, or even during a war if enough data has been collected during previous operations. But any guess about future usefulness of as yet nonexisting weapons is exactly that—a guess.

Finally, even after obtaining all the pertinent data, the decision makers have to consider the threat and the available data in light of other defense and general economic demands and establish priorities. The problem is one of both time and money and can be broken into three parts. First, is this a real threat, or alternately, does the new concept (or technology) answer a real need? Second, what should be done about it? Third, how can a suitable answer actually be developed? In many cases the third stage is the shortest one, although admittedly the most expensive one. The problem thus reverts to the decision makers and their ability to make timely decisions.

A high-ranking official of the United States Air Force said in 1998 that in the 1960s there was a lot of money for development and not enough technologies to be developed. Today there are a lot of technological choices but not enough money, so the choices are very hard to make. Today more than ever, "affordability" is becoming a key word. But it seems that without detracting from its importance, when used correctly, affordability now replaces other arguments as the sum and substance of old-fashioned conservatism and as a potential impediment to necessary, sometimes critical technological innovation. In fact, as noted in chapter 3, one of the arguments against the machine gun was its price and the price of its ammunition, in other words—affordability.

Volumes have been written about this topic, so there is no need to delve into it again here. In any case, it is beyond the scope of the present volume. Today, however, soldiers are already worried by the fact that their weapons were likely developed and produced by the lowest bidder. In addition, if in the future these weapons become suddenly obsolete or vulnerable because questions of affordability were wrongly answered fifteen or twenty years earlier, there is no doubt that some of the mistakes of World War I are apt to repeat themselves.

THE PROBLEM OF FEEDBACK

Technological warfare is an extremely complicated undertaking, with a lot of constantly changing parameters. Not everything can be predicted, but an integrated kind of thinking, with consideration of a multitude of potential scenarios is an absolute necessity. An important factor in the success of any such effort is the practice of feedback among the scientists, the engineers, the military, and the political decision makers.

Commercial technological efficiency, even in very complex or competitive fields, is easily measured by the profit and loss columns, and in the case of lagging performance usually can be recognized in time and steps taken to rectify the situation. What's more, even a total collapse, while most unpleasant to all involved, is rarely a matter of life and death. Military inefficiency or failure, which might be caused by technological failure, might be discovered too late, and the price of rectifying it, even if at all possible, is always casualties. Under the right circumstances, even small inefficiencies have a tendency to accumulate.

During World War II, the electronics research people in Britain established the so-called Sunday Soviets. These were meetings that brought together scientists, civil servants, high-ranking officers, and aircrew on operational postings. In these meetings, any technical matter could be raised and aired, and rank did not have any privileges. On occasion up to forty people were present, and the meetings were considered a huge success by bringing together the decision makers, the "suppliers," and the "clients" in an informal atmosphere. In the United States, many scientists were mobilized into research and development bodies. The OSRD, which financed much military research, generally had a very good working relationship with the military, although as pointed out earlier, some high-ranking officers were not always too happy with this arrangement and resented the encroachment of civilians into "purely military matters." In Israel, an even better arrangement was inadvertently achieved. Because of the all-encompassing reserve service, many of the scientists and the development specialists in the defense industries are also soldiers, spanning the whole range of ranks and up to the highest-ranking officers. Because of the regular training periods and the too-frequent fighting and wars, these people have an excellent grasp of both the operational needs of the services and the scientific and technological capabilities of the industry. Adding to this the very informal behavior in Israel, all these people conduct business on a first-name basis. Such traits simplify interpersonal communications, and as we have seen, on too many occasions poor interpersonal relations are the real stumbling blocks in military innovation. In a similar vein, it is interesting to note that of the original five members of the Tizard Commission (on air defense of Britain), all three scientists saw active service during the First World War. The quick and successful outcome of the deliberations of this committee might have had something to do with this prior service. These kinds of relationships would have been almost impossible during World War II in Germany and unimaginable in Japan.

Because of differences in national and organizational culture, the Israeli model is almost impossible to emulate in other countries. Training officers in the combat branches as scientists or engineers, although occasionally tried, will not work without major upheavals in thinking and practices in military forces. It seems that under the circumstances, the British World War II model is the best overall solution, even in peacetime. One thing is sure. Complacency and the NIH syndrome are slow-acting maladies, but once a certain threshold is reached, the results are disastrous and often quite sudden.

SOME CURRENT POTENTIAL CAUSES OF TECHNOLOGICAL FAILURES

From the above we can gather some insight as to where a technological failure starts. In most cases (though not all) it stems from anti-intellectualism, the inability to see the wider picture, and the inability to analyze more than one move at a time. The worst problem, however, is the inability to realize that the environment, both physical and abstract, is in a constant state of flux. In other words, the culprit is the "rule of conventional wisdom." If any of these actually happen and result in a technological failure, occasionally there is yet one more haven for those caught short—the ability to adapt quickly even under pressure, to dump preconceived notions overboard and look for ways to recover. While this is possible for individuals even within an atrophying system (consider the recovery of the Israeli Defense Force after the first days of the 1973 war even in the face of the Saggers and advanced antiair systems), this ability is rather rare, because a system usually is the sum total of its individual components. If the system really fails, it is likely that from the start there were not enough such individuals in the organization.

In the past, many "new" military innovations were actually the result of the marrying of several old and well-known technologies or whole weapons into something new, the usefulness of which was above the sum total of its components. This is called synergism. In the past, however, new inventions and concepts were based on fewer scientific disciplines, and both the number of technological innovations and the rate at which they were developed was smaller than it is today. As outlined in chapter 8, nowadays, scientific and technological developments in myriad fields have military applications. This proliferation of new scientific fields of learning harbors the potential for numerous permutations of various aggregates of such fields, possibly leading to a complete overthrow of our notions about "fighting a war." Furthermore, a huge scientific and technological establishment is no longer necessary for the development and exploitation of new military technologies. This situation is enhanced by two factors—the free and widespread dissemination of data by the electronic media, including the Internet, and the ease of purchase of both components and subsystems all over the world.

Information warfare is a case in point. It is based on the extensive use of sensors, computers, and communications technology in order to achieve one of two goals. The first goal is to glean all information on the enemy forces (and to prevent the enemy from obtaining such information on one's own forces) in order to conduct more efficient military operations. The second goal, when applicable, is to disrupt the enemy's "home front" by destroying or subverting its computers and everything it controls, including commerce and industry, communications and the mass media, leading to a possibly overwhelming collapse of an adversary even before the first shot has been fired. Because of the growing complexity of "weapon systems" (in their widest meaning) and difficulties in disseminating this information, the potential for a technological failure (and technological surprise) not only lurks in the shadows but also becomes larger with time.

LANCHESTER'S EQUATIONS AND THE EFFECT OF TECHNOLOGICAL INNOVATIONS

Finally, after extolling the virtues of technological innovation in military systems and its role in causing or preventing failure and surprise on the battlefield, and after having proved it at least empirically, can its value truly be proven in a mathematically rigorous way? To examine this question, it is useful to go back to the Lanchester equations. Let us assume (in the often-presented form of Lanchester's scenario) that all the infantry soldiers on one side (or just some of them) wear bulletproof vests. Without any regard to quantity or quality, in such a case the results of the classic Lanchester engagement will be preordained. This is not a totally imaginary or even a hypothetical case. Stealth aircraft do wear some sort of such protection.[12] In recent years some papers were published about "hard-kill" active defense (the destruction of incoming threats before they reach the target) for aircraft and submarines,[13] and some forms of "soft-kill" defense are operational on aircraft of various types.[14] Today these include chaff, flares, and jamming.

Imagine now the case of an air force introducing some device that will give its aircraft total immunity from all kinds of missiles. Can we incorporate such a "quality" (because obviously it is an enhanced quality) into the Lanchester equations, even assuming a one-on-one engagement? The difficulty lies in the fact that if such a device is novel or important enough to stand alone (and does not get incorporated into the existing overall measures of quality of the force), we cannot quantify a corresponding value for the aircraft of the other side. In fact, we cannot assign it any real value. Since temporarily, at least, the other side does not have the new device or technology, the value it has to receive for this particular measure of quality equals zero. In many calculations this might prove awkward, to say the least, but in this case it will imply an infinite, or at least an overwhelming superiority for the side owning the new device. Even if we bury such a device in the overall quality of the air force in question, it will show up in any calculation of attrition rates, which will become completely lopsided. A case in point is the introduction of the forward-firing machine guns on aircraft during World War I. The result of such a calculation is in fact the mathematical proof of the importance of new technologies on the battlefield.

All this is also true in the growing number of the so-called asymmetric kinds of warfare, mostly involving low-intensity conflicts between highly technological forces and less well-equipped ones, often including women and children. In many instances the less well-equipped forces use the inability or moral constraint of the better-equipped force to freely use its weapons. The development of the various nonlethal devices is one way of introducing new technology, to overcome the "advantages" enjoyed by the less-equipped force. Such an unexpected introduction of new technologies or devices of course works also for the "bad guys," from computer "hacking" to the introduction of chemical and biological agents into the midst of unprepared civilian populations.

By the nature of things such technological innovations do get introduced, by either side, and in most cases it is possible only to speculate what would have happened if the innovation had been introduced earlier in the game to have a real effect, but was not. Such speculation certainly is not an exact science. But in most cases it is vivid enough to make you feel either very lucky (if it was something examined or concocted, but not introduced, by the other side) or very foolish and angry if your side was the one that missed the opportunity.

NOTES

PROLOGUE. EARLY TECHNOLOGICAL FAILURE: THE BATTLE OF CRECY

1. Bernard and Fawn M. Brodie, *From Crossbow to H-Bomb* (Bloomington: Indiana University Press, 1973), 39.

2. Barbara W. Tuchman, *A Distant Mirror, The Calamitous 14th Century* (London: Macmillan, 1979), 71. The court people were afraid to inform the king of the disaster until the court jester was pushed forward and spoke about the cowardly English, who did not jump in the water like the brave French. The king got the point. It was also said that the fish drank so much French blood that if they could speak, they would have spoken French.

3. Archer Jones, *The Art of War in the Western World* (Urbana: University of Illinois Press, 1987), 157. Jones relates that there was some effort to introduce the longbow in France, but it failed because of the lengthy practice required to train an efficient shooter.

CHAPTER 1. PREDICTING THE OUTCOME OF CONFLICTS

1. On the subject of stupid misconduct of higher command see, for example, Norman Dixon, *On the Psychology of Military Incompetence* (New York: Basic Books, 1976).

2. The reader should note that in order not to use too cumbersome terminology the term "army" in this discussion encompasses all branches, including where applicable, the air force and the navy.

3. Geoffrey Regan, *The Guinness Book of Military Blunders* (Enfield, U.K.: Guinness Publishing, 1995), 131.

4. Sun Tzu was a Chinese philosopher and military leader who lived in the fifth century B.C. While his writing is more flowery than Clause-

witz's, he dealt with similar problems and, not surprisingly, came to basically similar conclusions. It is interesting to note that Sun Tzu was the first military writer who also professed that "political" considerations are important when planning a military campaign. This dictum served the Chinese Peoples' Army well, both in their fight against the Japanese and later against the Chinese National Army under Chiang Kai-shek.

5. Carl von Clausewitz, *On War* (New York: Barnes and Noble, 1966), 75.

6. Michael I. Handel, "Clausewitz in the Age of Technology," in *Clausewitz and Modern Strategy* (London: Frank Cass, 1986), 51–52.

7. A weighting function is a number or a fraction that will establish that one trait of quality, say training, is more important than (for example) resourcefulness, and by how much. These relative merits must be devised for the whole range of traits, and most important, they must be internally consistent.

CHAPTER 2. THE ROOTS OF TECHNOLOGICAL FAILURE

1. Clausewitz, Vol. I, 84.

2. Ibid., Vol. III, 250–251.

3. In the Crimean War (1853–1856), because of ambiguous orders a force of British cavalry charged along a valley lined on three sides with Russian artillery.

4. Regan, 83, 100.

5. See Elting E. Morison, "Gunfire at Sea: A Case Study of Innovation," in *Men, Machines, and Modern Times* (Cambridge, MA: MIT Press, 1966), 35.

6. Consider the Polish army, which before World War II devoted an inordinate percentage of its budget to horse cavalry, although admittedly supplying them with antitank guns. (See Steven J. Zaloga, "Polish Cavalry against the Panzers," *Armor Magazine* [January–February, 1984], 28.) But although there were many stories about Polish cavalry charging German tanks, Zaloga claims they were fabricated by both the Poles and the Germans, each for their own propaganda purposes. See Heinz Guderian, *Panzer Leader* (New York: Ballantine, 1961), 53; and Zaloga, 26 and 28.

7. John Keegan and Richard Holmes, *Soldiers, A History of Men in Battle* (London: Hamish Hamilton Ltd., 1985), 15.

8. David E. Johnson, *Fast Tanks and Heavy Bombers, Innovation in the U.S. Army, 1917–1945* (Ithaca, NY: Cornell University Press, 1998), 2.

9. J. F. C. Fuller, *Armament and History* (New York: Charles Scribner's Sons, 1945), 18.

10. It is related in several other places in the Bible that the Israelites were quite handy with the bow and arrow and the sling. See I Samuel 20:36 and Judges 20:16. Consequently, it is quite probable that the Philistine edict was instigated by economic factors. A whole people, being an agrarian society, had to come two or three times a year to the Philistine blacksmiths to have their tools fixed.

11. For example, the inventor of the firing cap was a Scot clergyman, Alexander Forsyth, who was an avid hunter and an amateur chemist. Richard J. Gatling (of machine gun fame) was trained as a doctor. Hiram Maxim, who finally perfected the machine gun (and who will be discussed later), was an artisan with no previous weapons experience.

12. A museum of da Vinci's technological and military works is in the Chateau Du Clos-Luce, in Amboise in the Loire Valley, France.

13. The "torpedo" in this case was a keg of gunpowder, placed near the target or allowed to drift toward it.

14. Wallace S. Hutcheon, *Robert Fulton* (Annapolis, MD: Naval Institute Press, 1981), 87. Pitt (William, the Young, 1759–1806) was the British prime minister at the time and supported Fulton's experiments.

15. Admittedly, the British navy came around after the Whitehead torpedo was perfected, and in 1871 were the first to establish a Whitehead licensed production facility.

16. Donald MacIntyre and Basil W. Bathe, *Man-of-War* (New York: Castle, 1974), 75.

17. Acetone is a critical chemical ingredient in the production of smokeless powder. Dr. Chaim Weizmann, a British chemist, converted an experimental process that he developed for the production of synthetic alcohol into a process to produce acetone, thus solving a major problem for the British war industry. Fritz Haber, who will be discussed in the next chapter, solved an even worse problem for Germany; he developed a process for the production of a cheap ammonia, an important precursor to the production of explosives.

18. David E. Johnson, 75.

19. Ibid, 4.

20. M. M. Postan, D. Hay, and J. D. Scott, *Design and Development of Weapons—Studies in Government and Industrial Organisation* (London: Her Majesty's Stationery Office and Longmans, Green and Co., 1964), 240.

21. Kenneth Macksey, *Military Errors of World War Two* (London: Cassell, 2000), 80.

22. Frank Whittle, *Jet* (London: Pan Books, 1957), 66–67.

23. Consider, for example, the engineer Cyrus Smith in Jules Verne's book *Mysterious Island*. The man was proficient in chemistry, explosives production, metallurgy, hydraulics, construction and demolition, astronomy, and navigation. And he did all these from memory, being cast with his companions onto an uninhabited Pacific island. While admittedly a fictional character, this was not much above the average for the period, the middle of the nineteenth century. Today these kinds of accomplishments often require half a dozen specialists, all armed with laptop computers.

24. Brodie, 234.

25. Robert Buderi, *The Invention That Changed the World* (New York: Simon & Schuster, 1997), 158. All told, the argument turned on about three dozen airplanes.

26. Ibid., 161. This kind of myopic reasoning was prevalent not only during World War II or only in the United States or its naval service. See also Ernst Udet's reaction to radar, described at the end of this chapter.

27. Ibid., 159.

28. Ibid., 227.

29. Macksey, 115.

30. Apart from their own efforts in this field, the Soviets received large numbers of aircraft from the United States through the "lend lease" program. Their favorite in this field was the Bell P-39 Airacobra. In the U.S. and England it was considered at best a mediocre fighter airplane, but the Russians loved it because it carried a very big (thirty-seven-millimeter) gun, and thus was an excellent tank buster.

31. The first real scientist who considered and published in this field was Constantin E. Tziolkovsky (1857–1935), who in 1904 published a technically sound monograph titled "Space Exploration by Means of Reaction Propulsion Craft," about achieving spaceflight by means of rockets.

32. Anthony J. Watts, *The Royal Navy, An Illustrated History* (Annapolis, MD: Naval Institute Press, 1994), 120.

33. Winston S. Churchill, *The Hinge of Fate* (Boston: Houghton Mifflin, 1949), 126; and *Closing the Ring* (Boston: Houghton Mifflin, 1949), 10.

34. Assembling a certain number of ships in a convoy reduces the probability of the convoy's detection by the same number, and the size of

the convoy has practically no bearing on the problem. It is true that a convoy's speed is limited to the speed of its slowest ship, and once detected there are many targets for submarines; but on the other hand, a convoy can be provided with escorts, while individual ships cannot.

35. ASDIC was actually the acronym for the Allied Submarine Detection Investigating Committee but was also accepted as the generic name of that device. On the other hand, Robert Gannon quotes a source that claims no such committee ever existed and the name was derived from "Anti Submarine Division-ics" (or devices). (See Robert Gannon, *Hellions of the Deep* [College Park: Pennsylvania State University Press, 1996], 60.) Later the term ASDIC was replaced by SONAR—SOund NAvigation Ranging.

36. Holger Herwig, in Williamson Murray and Allan R. Millett, eds. *Military Innovation in the Interwar Period* (Cambridge, UK: Cambridge University Press, 1996), 251.

37. There were some noted exceptions, which will be discussed later, in chapter 6.

38. Allan R. Millet and Williamson Murray, *Military Effectiveness*, Vol. 1, The First World War (Boston: Unwin Hyman, 1988), 84–85. von Falkenhayn later replaced Moltke as chief of staff.

39. Basil H. Liddell Hart, *Foch, The Man of Orleans* (Boston: Little, Brown, 1932), 44. Admittedly, a decade after the war, Foch changed his mind and stated that "air attack, by its crushing moral effect on a nation, may impress public opinion to the point of disarming the government and thus become decisive." Ibid, 44. This was essentially the approach used by Guilio Douhet.

40. Brodie, 110.

41. Hutcheon, 60.

42. Buderi, 147–148.

43. David Pritchard, *The Radar War* (Wellingborough, UK: Patrick Stephens Ltd., 1989), 64.

CHAPTER 3. MISUNDERSTANDING THE BATTLEFIELD ENVIRONMENT

1. John Ellis, *The Social History of the Machine Gun* (New York: Arno, 1981), 34.

2. To be fair, it should be pointed out that there were officers, includ-

ing high-ranking ones in the British army, who thought favorably of machine guns and even said and wrote so. However, they were a minority and their voice was drowned in the general air of rejection of these weapons.

3. Donald R. Morris, *The Washing of the Spears* (New York: Simon & Schuster, 1965, fourth printing), 371–373. They first met disaster at Isandhlwana in January 1879, when a force of some seventeen hundred troops—about half of them British and the rest native troops—was almost totally wiped out by a force of more than twenty thousand Zulu warriors. The cause of the debacle can be traced to the refusal of the quartermasters to break open ammunition boxes during the engagement. They insisted on opening them in a proper manner, but it was a slow and tedious process. They also refused to give ammunition to native runners not from their own command. Some sections of the firing line finally ran out of ammunition and were rushed by the numerically superior Zulus.

4. This was the battle that finished the Mahdi and the Dervishes uprising in the Sudan. While the British later made much of the "Thin Red Line," which undeniably was pretty thin, it nevertheless was quite thickened by several machine guns. The final tally was eleven thousand Dervishes dead to forty-eight British and Egyptian killed. (Total casualties were about fifteen thousand to five hundred British and Egyptian.)

5. From English poet and author Hilaire Belloc's "The Modern Traveller," taken from Ellis, 18.

6. Liddell Hart, 31.

7. The numbers in the table were taken from John Laffin, *Brassey's Battles* (London: Brassey's, 1995).

8. Ibid., 463.

9. Included in the figures for the French are about 1.1 percent of Belgian casualties.

10. It was unfortunate that the two battles that conclusively demonstrated the power of the breech-loading rifle were fought after and not before Ripley took office, although it is questionable if it would have helped. The first was the Battle of Gettysburg, July 1–3, 1863. On the first day, a large force of Confederates (seven thousand strong) ran into twenty-five hundred dismounted Union cavalry, armed with single-shot breech-loading Sharps and seven-shot Spencer rifles. The Confederates were surprised by the effects of the superior firepower and badly mauled. The second was the Battle of Sadowa, described in chapter 1.

11. It was suggested that Custer had not digested the fact that his

troopers' firepower was enormously diminished when he rode to his last battle.

12. Clint Johnson, *Civil War Blunders* (Winston-Salem, NC: John F. Blair, 1997), 51.

13. Ellis, 74.

14. Bruce A. Rosenberg, *Custer and the Epic of Defeat* (University Park: Pennsylvania State University Press, 1974), 9, 25.

15. It should be remembered that the Battle of Little Big Horn had another adjunct. Major Marcus Reno and Captain Frederick Benteen, with nearly another four hundred troopers of the 7th Cavalry, were also surrounded by Indians on that sweep. They held off the Indians for more than a day, with the loss of fifty-three killed, until the Indians retreated. Could it be possible that if armed with a better rifle, they could have broken out and saved some of Custer's men?

16. H. H. Arnold, *Global Mission* (Blue Ridge Summit, PA: TAB Books, Military Classics Series, 1989), 38.

17. It is fairly easy to convert artificial fertilizers into low-power explosives and vice versa; explosives treated in a certain manner can be used as fertilizers.

18. There were in fact two solutions for this problem, but both required prodigious amounts of electrical power.

19. For parts of the following narrative, I am indebted to Joseph Borkin, *The Crime and Punishment of I. G. Farben* (New York: The Free Press, 1978).

20. The announcement of the prize was accompanied by a major scandal when his role in the German chemical warfare effort (see below) was publicly aired and he was almost tried as a war criminal together with nine hundred other Germans, including the kaiser. Due to mismanagement on the part of the Allies, the list was reduced to some forty-five obscure persons who got what amounted to ridiculous sentences. Haber was not one of them, and the road was cleared for him to receive the prize.

21. Bosch eventually became the general manager of I. G. Farben, and in 1931 (together with Friedrich Bergius) he too won the Nobel Prize in chemistry for work on chemical high-pressure systems.

22. Rathenau was considered a brilliant intellectual and also had political connections. After the war, he served as Germany's foreign minister. It is interesting to compare Rathenau's position, deeds, and clashes with his country's military establishment with those of Vannevar Bush, the head of the American OSRD under President Roosevelt. See chapter 6 in this volume.

23. There still was some sporadic movement of German ships in and out of Chile. Finally, under pressure from the Allies, Chile interned all the German ships in the Valparaiso harbor, about a hundred of them, for the duration. It is interesting to note that nearly two-thirds of these ships were sailing ships. See Anon., "Los Buques Alemanes." Admittedly, bulk nitrates can very well be carried in such inefficient ships, and their upkeep was probably low. But one cannot avoid the comparison with the German army's use of horse-drawn wagons to supply the panzer divisions on the eastern front during World War II.

24. It seems that the name Admiral Graf Spee bode ill fortune for German naval operations in the South Atlantic. On December 13, 1939, a British squadron intercepted the "pocket battleship" (in essence a cruiser with bigger guns) *Admiral Graf Spee* and after a short fight, the Battle of River Plate, forced her to seek shelter in the harbor of Montevideo, where she was later scuttled by her crew. Her captain committed suicide on the following day. See also chapter 6, regarding Henry Tizard and the Battle of the Beams.

25. Churchill, *The World Crisis, 1911–1914*, 474.

26. S. S. Swords, *Technical History of the Beginnings of Radar* (London: Peter Peregrinus Ltd., 1986), 120–144. Swords also lists Hungary, which joined the radar club during the war, but on her own, without outside help.

27. In 1922 two scientists, Albert Hoyt Taylor and Leo C. Young, working for the navy, realized the possibility of a radio detection system. They wrote a letter to the Bureau of Engineering, but it was not even answered, and so they dropped the whole matter until 1930, when they would pick it up again. See David K. Allison, *New Eye for the Navy: The Origin of Radar at the Naval Research Laboratory* (Washington, DC: Naval Research Laboratory, NRL Report 8466, 1981), 40–41. Allison, however, makes a curious comment, saying maybe it was better that the two scientists never got a reply to their first letter. Since the technology was immature, even if a working model had been built it would have been crude and cumbersome, and this might have turned the navy away, and later slowed development. Luckily for all, the British did not have such inhibitions in 1935, but the practical question remains: when does a new technology become mature enough, and who is to decide?

28. The familiar television and radar Yagi antenna was in fact invented by a Japanese man by the name of Hidetsugu Yagi.

29. RDF stands for Radio Direction Finding. The name RADAR (RAdio Direction And Ranging) came into use later during the war.

30. Peter Townsend, *Duel of Eagles* (New York: Simon & Schuster, 1970), 169, 172.

31. This is noteworthy in view of previous comments on the importance of speedy communications.

32. Townsend, 333.

33. Bernard L. Boylan, "The Development of the American Long-Range Escort Fighter" (Ph.D. diss., University of Missouri, 1955), 10.

34. Later, in June 1935, Baldwin again became prime minister. He started the British rearmament program and was finally replaced, in May 1937, by Neville Chamberlain, of Munich infamy.

35. *New York Times*, November 11, 1932, 4.

36. The Me-109 is sometimes identified as Bf-109, standing for *Bayerische Fleugzeugwerke* (Bavarian Aircraft works), where the Messerschmidt main plant was located.

37. The .50-caliber machine gun, with minor modifications, is still in use in many armies in the world.

38. William R. Emerson, *Operation Pointblank (A Tale of Bombers and Fighters)* (U.S. Air Force Academy, CO: The Harmon Memorial Lectures in Military History, 1962), 24.

39. This was the famous "Dam Busters" raid. The Ruhr valley's steel industry used some two thousand gallons of water for each ton of steel produced. The Ruhr's water supply was regulated by several dams. The dams were protected by antitorpedo netting, and there was no way to hit them accurately from the air. A British engineer, Barnes Wallis, devised a specially configured bomb that was to be dropped at a distance from the dam and then hop and skip on the water's surface until it hit the dam, sank to its base, and there exploded by hydrostatic pressure. Initially this scheme met with considerable derision, because the special bombs were to weigh five tons and "skipping" five-ton "pebbles" was considered impossible. But when the dams suddenly became a priority target, Air Chief Marshal Arthur Harris, head of Bomber Command, accepted the idea. In May 1943 the raid, by the specially created 617 Squadron, was carried out with considerable success. Two dams were destroyed and a third damaged at the price of eight lost bombers out of nineteen that set out. The water supply to the Ruhr industries was disrupted for a while and a lot of peripheral damage was caused. The 617 Squadron remained on the rolls and was later used for high-accuracy missions, including the dropping of the five- and ten-ton "Earthquake Bombs."

40. Franklin D'Olier, Chairman, *The United States Strategic Bombing Survey, Summary Report, (European War)*, September 1945, 14.

41. The second-generation missiles such as the American TOW (Tube-Launched Optically Guided Wire-Controlled) and the French HOT (High-Subsonic Optically Tracked Tube-Launched) improved on this system. The operator has only to point the sights at the target, and the guidance system (which knows where the target is and the position of the missile relative to it at any given moment) will make the missile follow the line of sight to the target. This is called SACLOS, Semi-Automatic Command to Line of Sight.

42. It could be interesting to speculate what would have happened if the Israeli creativity in weapon systems was channeled earlier to development of antitank missile systems.

43. It is true that today practically all of the infantry ride in so-called APCs (armored personnel carriers), but these are essentially only armored buses that bring the infantry to where the action is. The infantry must then dismount and fight on foot.

44. On the same date the Syrians also attacked, but that part of the war does not directly involve this story. Although the Syrians were also equipped with Saggers, surprisingly they made practically no use of them.

45. It was still difficult to achieve a hit—on the average nearly fifteen missiles were required to score one.

46. W. Seth Carus, testimony in "Report of the Commission to Assess the Ballistic Missile Threat to the United States," Appendix III: Unclassified Working Papers, pp. 79–85, July 15, 1998.

47. Among others, by Lieutenant Colonel Vladimir (Popsky) Peniakoff, DSO, a maverick British officer who established and later commanded a successful intelligence and raiding unit that operated in the Western Desert and later in Italy.

CHAPTER 4. MISUNDERSTANDING AVAILABLE TECHNOLOGY

1. Winston S. Churchill, *The World Crisis 1915* (New York: Charles Scribner's Sons, 1923), 72–77.

2. Ronald W. Clark, *War Winners* (London: Sidgwick & Jackson, 1979), 20.

3. There is much information about early tank development, but I chose to rely mostly on Churchill's history of the First World War. See Churchill, *World Crisis 1911–1914*, 342–345, and *World Crisis 1915*, 61–83.

4. Since people are often influenced for life by early combat experiences, it is probable that Churchill remembered that episode well, even though in this case he was on the giving side.

5. Churchill, *World Crisis 1915*, 65.

6. Churchill was the one who before the First World War finally pushed England and the Royal Navy into the oil age, by involving the British government with the Anglo-Persian Oil Company. The conversion to oil was started by Admiral John Fisher, and Churchill made it total. See Daniel Yergin, *The Prize* (New York: Simon & Schuster, 1992), 150–164.

7. The part of the reply about the difficulties of travel over enemy-held ground shows that he really did not have the foggiest idea as to what was proposed, since the whole point was to nullify those difficulties, especially machine gun fire.

8. Churchill, *World Crisis 1915*, 67.

9. The Gallipoli operation was an Allied landing in the Dardanelles in Turkey in April 1915, carried out to force the Ottoman Empire out of the war. The Allies managed to secure a beachhead, but poor organization, inept leadership, and basic tactical errors (on two occasions, after landing unopposed, instead of pushing inland the troops dawdled for hours on the beach until the Turks arrived) finally forced an evacuation in January 1916.

10. Churchill, *World Crisis 1915*, 69.

11. This is because when size, say, doubles, the weight of the airframe grows a little more than twice; the aerodynamic drag is approximately two squared (four times the original), but the volume grows to eight times the original. Even after considering the weight of the bigger engines to overcome the increased drag and their increased fuel consumption, the net increase in payload-carrying capacity is quite substantial. But to be effective, bombers must carry a considerable weight of bombs and ammunition for the defensive guns, while fighters must carry only the ammunition for their guns. If proper attention is paid to the aerodynamic design, fighters can equal, and even surpass, the range of fully loaded bombers.

12. Cajus Bekker, *The Luftwaffe War Diaries* (New York: Ballantine, 1971), 236.

13. Ibid., 217.

14. Benjamin S. Kelsey, *The Dragon's Teeth?* (Washington, DC: Smithsonian Institution Press, 1982), 66.

15. Adolf Galland, *The First and the Last* (New York: Ballantine, 1954), 24.

16. Alexander de Seversky, *Victory through Air Power* (New York: Simon & Schuster, 1942), 67; and Galland, 30–31.

17. Murray and Millett, eds., "Strategic Bombing," in *Military Innovation*, 133.

18. de Seversky, 53–56, 199–201.

19. During World War I the British developed the "K" class of steam-powered submarines. They were extremely fast, but proved too complex and disaster prone and were eventually discarded. See Don K. Everitt, *K-Boats, Steam-Powered Submarines in World War I* (Annapolis, MD: Naval Institute Press, 1999).

20. A submarine is slow and unarmored, and its main firepower, the torpedo, is too specialized as it is designed to attack only other water-borne vessels.

21. Charles M. Sternhell and Alan M. Thorndike, *Antisubmarine Warfare in World War Two*, OEG Report No. 51, Operations Evaluation Group, Office of the CNO (Washington, DC: 1946), 58–59.

22. MacIntyre and Bathe, 252.

23. Bill Gunston, *Submarines in Color* (New York: Arco, 1977), 116.

24. Hutcheon, 44.

25. Bill Gunston, 116.

26. Although based on different propellants, these systems are now called AIP, for Air-Independent Propulsion.

27. The U.S. Navy had considered the use of hydrogen peroxide as a component of torpedo fuel since 1930, and it was based on an old American patent from 1915. Interest was renewed again in 1940, and it was shown that it had superior performance and no wake, but again the chemical (called NAVOL for Navy Ordnance Laboratory) was considered too dangerous to introduce into submarines. Eventually a torpedo so powered, the Mark-16, was produced but too late to see combat. Its production was continued, though, for another thirty years.

28. Some of them even had a deck storage locker for a small, folding gyrocopter, a type of rotating-wing aircraft, which could be towed aloft to increase the observation range.

29. Rommel's forces captured an antisubmarine-radar-equipped Wellington in Tunisia, and he actually impressed it into service for detecting Allied ships. See E. G. Bowen, *Radar Days* (Bristol, UK: Adam Hilger, 1987), 113–114. It was several months before Berlin found out about this and demanded the airplane for investigation.

30. Boylan, 33.

31. Kelsey, 65.

32. Murray and Millet, *Military Innovation*, 120.

33. Williamson Murray, *The Luftwaffe, 1933–1945: Strategy for Defeat* (Washington, DC: Brassey's, 1996), 224.

34. Boylan, 206.

35. Ibid., 54–55.

36. Emerson, 29–30.

37. Boylan, 156.

38. Arnold, 495–496.

39. James P. Baxter, *Scientists against Time* (Boston: Little, Brown, 1946), 198–199.

40. Commission on the Organization of the Government for the Conduct of Foreign Policy, June 1975, *Adequacy of Current Organization: Defense and Arms Control*, Vol. 4, Appendix K, 192.

41. Ibid., 194.

42. Ibid., 195.

43. Ibid., 197.

CHAPTER 5. BAD MANAGEMENT, WORSE LEADERSHIP, AND NIH

1. The description of the development of the Leigh Light is partially based on Alfred Price, *Aircraft versus Submarine* (London: Jane's, 1980), 59–65.

2. It is quite possible that these calculations the Germans made were based on Scottish physicist James Maxwell's (1831–1879) original work and on an extension of that work performed by distinguished German physicist Arnold (Johannes Wilhelm) Sommerfeld (1868–1951). For a brief history of modern stealth development, see Ben R. Rich and Leo Janos, *Skunk Works* (Boston: Little, Brown, 1994), 19–20.

3. Bowen, 114.

4. Price, *Aircraft versus Submarine*, 115–116; and David E. Fisher, *A Race on the Edge of Time* (New York: McGraw-Hill, 1988), 306.

5. Buderi, 158.

6. Richard Overy, *Why the Allies Won* (New York: W. W. Norton, 1995), 60.

7. Price, *Aircraft versus Submarine*, 118.

8. Sternhell and Thorndike, 58.

9. Admiral Scott was a very forward-thinking officer. In June 1914 he caused a minor furor when in a letter to *The Times* he predicted that submarines and aircraft would make the battleships useless and advocated a naval policy based on a large air force, a large fleet of submarines, and cruisers for trade protection. See Paul M. Kennedy, *The Rise and Fall of British Naval Mastery* (London: Ashfield Press, 1983), 199.

10. E. Morison, "Gunfire at Sea," 31.

11. Price, *Aircraft versus Submarine*, 62.

12. The story of the plastic armor is based on Gerald Pawle, *Secret Weapons of World War II* (New York: Ballantine, 1968), 62–70.

13. The DMWD was the British Admiralty's "skunk works" department. They were presented with operational problems and were tasked with finding quick solutions, without going through the standard and lengthy process of development. By virtue of comprising civilian scientists and specialists in many fields, mobilized for the duration, they usually were very quick in finding solutions, although naturally not all were successful. Also, coming from the civilian sector, the DMWD had good connections and could draw on a wider circle of specialists. One of their developments was the adaptation of an airfield protection device, the PAC (Parachute and Cable) for use on small merchant vessels. It was a small rocket that was fired from the deck of the ship, dragging a cable with a parachute, into the path of an attacking airplane, and that claimed several German aircraft. Another DMWD development was the "hedgehog," a pattern of improved depth charges shot simultaneously forward from the sub hunter, that was considered a tremendous improvement over the then-standard system of rolling the depth charges from the stern of the ship.

14. The landing at Dieppe, in August 1942, was a test of ways and means to take a defended beach by means of seaborne assault. About six thousand British, Canadian, and American troops, accompanied by tanks and about 250 landing boats and other ships tried to take the town and its port. Because of superficial intelligence gathering, and poorer planning, the operation failed with some 70 percent casualties. The lessons, however, were learned and successfully applied in later operations.

15. G. Macleod Ross, *The Business of Tanks, 1933 to 1945* (Ilfracombe, UK: Arthur H. Stockwell Ltd., 1976), 38.

16. Ibid., 87.

17. Macksey, 84.

18. A discarding sabot shell was a German invention in which a core of

heavier metal—steel or tungsten (and today even depleted uranium)—is surrounded by an aluminum shell. This aluminum shell is discarded when the round emerges from the muzzle, and only the inner core flies on. Overall, such a shell is lighter than a solid shell and thus has a better acceleration in the barrel. Because of the gun's internal ballistic considerations, the larger diameter enables higher acceleration in the barrel, while the smaller diameter of the core reduces drag during flight. While today, under certain circumstances, a smaller-diameter core has better penetration, at the time, and with the armor then in use, the diameter did not really matter.

19. More than fifty thousand M-4s were produced.

20. Peter S. Rosen, *Winning the Next War* (Ithaca, NY: Cornell University Press, 1991), 188.

21. Ibid., 188. It should be pointed out that after America's entry into the war a somewhat better gun, a seventy-six-millimeter, was developed, but this one too had a muzzle velocity of only 793 meters per second, which left a lot to be desired. Only much later was a "sub-caliber" with a muzzle velocity of over one thousand meters per second developed for it.

22. Omar N. Bradley, *A Soldier's Story* (New York: Henry Holt and Company, 1951), 322–323.

23. Lida Mayo, "The Ordnance Department: On Beachhead and Battlefront," in *United States Army in World War II—The Technical Services* (Washington, DC: Office of the Chief of Military History, United States Army, 1968), 325.

24. Ibid., 327.

25. Albert Speer, *Inside the Third Reich* (New York: Macmillan, 1970), 233; and Shlomo Na'aman and Roni Cohen, "Tank Armament, Past, Present and Future," in *Ma'arachot* No. 247–248 (1975), 32.

26. Ross, 283, 300, 311.

27. Ibid., 283, 300.

28. Ibid., 41.

29. Ibid., 301.

30. Ibid., 316.

31. The following account is largely based on Edwyn Gray, *The Devil's Device* (Annapolis, MD: Naval Institute Press, 1991), 222–224; and Gannon, 74–93.

32. Gannon, 81.

33. Clay Blair Jr., *Silent Victory* (Philadelphia: J. B. Lippincott, 1975), 439.

34. Gannon, 85–86.
35. Ibid., 86.
36. Macksey, 113.
37. Gray, 223.
38. Blair, 879.
39. See Herwig, in Murray and Millett, *Military Innovation*, 260.
40. Blair, 20.
41. Galland, 159–160.
42. James P. O'Donnell, "The Secret Fight That Doomed the Luftwaffe," *Saturday Evening Post*, 8 April 1950, 103.
43. They reorganized their night fighters' control system, positioned the fighters to use the searchlights, and even the clouds, to better see the bombers and developed radars that were immune to the then-current window. Not all solutions worked at all times, but the Germans were fairly good at adapting to changing circumstances. The problem was that most of the time they were reacting to new Allied measures and initiatives, and on occasion the learning curve was very shallow. In other words, it took them too long to develop and introduce a good solution, and by that time the Allies usually had arrived at something better.
44. Even radar approaches were good only to about three hundred meters. After that distance, the radar became ineffectual and the final approach had to be made visually.
45. Bekker, 501–502.
46. This episode, and the following one about escape hatches, are related by Dyson Freeman, *Disturbing the Universe* (New York: Harper & Row, 1979), 23–28.
47. See Air Ministry, *The Origins and Development of Operations Research in the Royal Air Force* (London: Air Ministry, Air Publication 3368, 1963), 66. These numbers differ somewhat from the ones quoted by Freeman—15 percent for Lancasters and 25 percent for the Stirlings and Halifaxes.
48. Ibid., 66.
49. Freeman, 28.
50. See R. V. Jones, *The Wizard War* (New York: Coward, McCann & Geoghegan, 1978), 466.
51. Fisher, 307.
52. Rich and Janos, 162–163.
53. Thomas Powers, *Heisenberg's War* (Boston: Little, Brown, 1993), 449.
54. David Irving, *The German Atomic Bomb* (New York: Simon & Schuster, 1967), 84–86.

55. Jeremy Bernstein and David Cassidy, "Bomb Apologetics: Farm Hall, August 1945," *Physics Today* 48, no. 8, part 1 (August 1995). Also see Leslie Groves, *Now It Can Be Told* (New York: Harper & Row, 1962), 333–335.

56. The following account is based to a large extent on Richard Rhodes, *The Making of the Atomic Bomb* (New York: Penguin, 1988).

57. Pacific War Research Society, ed., *Japan's Longest Day* (New York: Ballantine, 1983), 43, 48, 58.

CHAPTER 6. PRECONCEIVED IDEAS, OVERCONFIDENCE, AND ARROGANCE

1. This of course is not always so and depends no less on the outlook of the superior and the general atmosphere in the institution in question.

2. See Jones, 93; Swords, 120–144.

3. And this was some of the top leadership of a country in a desperate war, shortly after the Dunkirk Evacuation.

4. A solid-fuel rocket is filled with propellant and is basically a big combustion chamber that has to withstand an internal pressure of about thirty to forty atmospheres (450 to 600 psi). This requires a robust construction that translates into weight. Consequently, solid-rocket motors are usually rather slender, a fact that enables reducing the thickness, and the weight, of the motor's walls.

5. In a liquid-fueled rocket, only a small combustion chamber is necessary, while the fuel tanks have only to withstand their own weight and maneuvering stresses.

6. There was one odd case where phones actually served better. In June 1940 General Heinz Guderian, the noted German armor commander, was at the head of an armor corps that was advancing very rapidly in France. His superior, General Ewald von Kleist, the commander of the armor group, was worried about this unexpected success and ordered Guderian to slow down, but Guderian paid no heed. There was a bitter confrontation, and Guderian tendered his resignation on the spot. Cooler heads prevailed, in particular Field Marshal Gerd von Rundstedt, the commander of the army group, and a compromise was achieved according to which Guderian's headquarters would stay in place and only reconnaissance units would advance. Guderian had his signal men lay down miles and miles of telephone lines to prevent his superiors from

listening to his orders to his forward units, and he felt again free to unleash his panzers.

7. The technology of directional antennas was not applicable since it is useful only if the location of both stations is precisely known and they are essentially stationary, and under certain circumstances the transmissions can still be listened to.

8. Herodotus (who lived in the fifth century B.C.) relates a story that a message was tattooed onto the shaved head of a slave. When his hair grew back, he was sent to the recipient, who had him shaved again.

9. Richard M. Ketchum, *The Borrowed Years 1938–1941* (New York: Random House, 1989), 735–737, 753.

10. Excellent descriptions and explanations of the internal working of the machine are given in Brian Johnson, *The Secret War* (New York: Methuen, 1978); David Kahn, *The Code Breakers* (New York: Scribner, 1996); and Gordon Welchman, *The Hut Six Story* (New York: McGraw-Hill, 1982).

11. These three countries had a defense pact. World War II actually started after Germany attacked Poland and Britain and France declared war on Germany.

12. Kahn, 503; see also Price, *Aircraft versus Submarine*, 125.

13. To some extent, at least, this is obvious. What one man can devise, another man can figure out. The Germans for a long time were reading the British merchant marine codes. This enabled them to deduce the positions of the Atlantic convoys and attack them.

14. The Japanese diplomatic service used a machine, adapted from the German Enigma, which was named Purple by the United States. The colors were assigned according to the complexity of the cipher, with the darker colors denoting more secure ciphers. The Japanese navy (and army) used codes, most of which were further enciphered. The effort to solve the Purple traffic was named Magic, while the British project to read the Enigma was named Ultra.

15. It seems that the Japanese had a predilection for this kind of arrogant thinking. As mentioned above (chapter 5, the Japanese Atomic Bomb), based probably on their own inability to do so, they concluded that the United States did not have the industrial capacity to pursue such an undertaking. Also, the island-hopping strategy of the Americans was initially excluded from Japanese thinking. This strategy required enormous quantities of naval craft that they were incapable of producing, and thus they concluded that the United States would also be unable to do so. Macksey, 215.

16. Kahn, 26.

17. Buderi, 34.

18. Vannevar Bush, *Modern Arms and Free Men* (New York: Simon & Schuster, 1949).

19. The book was published in 1949 and written presumably a year or two earlier.

20. Bush, 52–53.

21. Ibid., 45.

22. Ibid., 44.

23. Ibid., 47.

24. Ibid., 100. In the early sixties, turn radii of five hundred to six hundred meters were achieved. A current F-16 can turn with a radius of three hundred meters at high subsonic speeds. Such a statement from a former chairman of NACA, which was at the cutting edge of technology in the thirties and forties, is rather puzzling.

25. After Korea the U.S. Navy and Air Force neglected dogfighting, until shocked into wakefulness by losses in the sky over Vietnam. Both services then instituted measures to correct this deficiency. See Curtis Peebles, *Dark Eagles* (Novato, CA: Presidio, 1995), 217–219, 231. Part of the problem was that missile-equipped aircraft were intended to shoot down only bombers and transport aircraft. Could Bush's ideas have influenced the fighter community?

26. Bush states (85) that chemical fuels had reached their zenith of performance. The fuel of the V-2 (liquid oxygen and alcohol) had an Isp (a measure of rocket performance with a particular fuel, representing pounds of thrust achievable from the flow of one pound of fuel per second) of 300. Better liquid fuel combinations with Isp of 370 were developed already in the sixties. The space shuttle's main engines, running on liquid oxygen and liquid hydrogen, have an Isp of 450, 50 percent more than the V-2. In any case, the problem is academic. Except for certain third-world developments, such as in Iraq and North Korea, which were pursued because of convenience and not because of efficiency, most current military missiles are based on solid fuel, which has considerably less Isp but other advantages.

27. Bush, 85.

28. Ibid., 86. He is referring to a range of approximately three thousand miles. As a rule of thumb, the maximum height of the trajectory of a ballistic missile is about 25 percent to 35 percent of the range.

29. Ibid. Today some intercontinental ballistic missiles are accurate to within several hundred feet.

30. Ibid., 85.

31. Ibid., 105–106.

32. Ibid., 93. Here Bush probably voices the then-accepted estimates of General Leslie Groves, the head of the Manhattan Project, and of J. R. Oppenheimer, its scientific leader.

33. Robert S. Norris, "Estimated U.S. and Soviet/Russian Nuclear Stockpiles, 1945–1994," *The Bulletin of the Atomic Scientists* (November/December 1994). I chose to go only up until 1957, the year the Soviet Union launched the first Sputnik, and in essence confirmed that it has a missile that can carry a nuclear warhead to any place on the face of the earth.

34. Bush, 94.

35. Overy, 180–190.

36. Emerson, 41.

37. Bush, 99.

38. Ibid., 83. One wonders in this connection if Bush was aware of the ineptitude of the U.S. Navy's BuOrd in fixing the torpedo problems.

39. Buderi, 28.

40. Bush, 106.

41. See Chuck Hansen, *U.S. Nuclear Weapons* (Arlington, TX: Aerofax, 1988), 171–176. A small weapon fired from a recoilless gun, the "Davy Crockett," was also developed in 1961. See Hansen, 198. While one might question the tactical use of such a weapon, it still was a successful development. Furthermore, an air-to-air missile, with a nuclear warhead was also developed.

42. Bush, 108.

43. Ibid., 98, 100.

44. Ibid., 87, 100. Bush mistakenly differentiated between jet and turbojet engines, as if the turbojet was an advanced form of the jet.

45. Ibid., 83.

46. Buderi, 156.

47. Concerning the width of the German radar capabilities, see Buderi, 202.

48. In the acknowledgments section of his book, Bush lists several high-ranking persons, both in the administration and in the sciences (including James Conant) who read the book and helped Bush. There is, though, the usual disclaimer that the opinions expressed are not to be interpreted as held by the people listed.

49. Buderi, 405.

50. The missile gap never really existed, and in a way it came into "being" because of lack of intelligence on Soviet intentions. The Soviets had a certain potential to produce missiles, which they never fully applied, and the Americans made an enormous effort to redress this imaginary inferiority. McNamara admitted to this in a public speech in 1967. See Rosen, 218–219.

51. Bush, 73. One wonders what Bush would have thought of fractals or fuzzy logic and their current applications.

52. A Delphi poll is a technique to decide questions in science policy (and is applicable, of course, to other fields) where a balanced opinion of professionals is sought. It consists of a questionnaire dealing with the subject and answered by the people polled. Based on the results of the first poll, a second questionnaire is prepared and distributed to the same group, enabling them to modify their opinion in light of new findings. Analysis of the individual answers and of the overlap in the reasoning is purported to give better-than-average predictions. The process can be done by a paper questionnaire or by computer in a "real-time" conference.

53. Note also the case of Admiral Jervis St. Vincent (see chapter 2), who turned down Robert Fulton and his "torpedo."

CHAPTER 7. POLITICAL AND IDEOLOGICAL MEDDLING

1. See Overy, 1–24 and 180–244; and Christopher Chant, ed., *Warfare in the Third Reich* (New York: Smithmark, 1996), 86–101.

2. Even the Battle of the Bulge does not disprove this argument. While certain initial successes were impressive, aided by Allied complacency, the whole effort literally fizzled out for lack of fuel and real airpower.

3. Until 1939 there were thirty-four Germans, twenty-three British, sixteen French, and fourteen American Nobel laureates.

4. Rhodes, 184–185.

5. This happy state of events can occasionally backfire, as happened with the stealth concept. This technology was initially developed in the United States from an innocuous set of mathematical equations published by a Russian scientist in a Russian conference. See also chapter 5, Note 2.

6. Macksey, 157–158.

7. This sort of mismanagement is quantitatively meaningless by itself

because although it lowers efficiency, it does not totally destroy the scientific effort, and lower efficiency in research is hard to quantify. But the Germans' mismanagement should be compared against the Allies' treatment of the same problem. Even before the war started, England had embarked on a systematic canvassing of its scientific manpower. The United States deferred the draft of many graduate students and scientists, who then contributed to the war effort in many fields.

8. Gannon, 202.

9. The Heisenberg Principle—"The Indeterminancy Principle of Quantum Mechanics"—states that it is impossible to measure simultaneously both the velocity and the position of atomic particles.

10. Galland, 163–164; and Johannes Steinhoff, *The Final Hours* (Baltimore, MD: Nautical & Aviation Publishing Company of America, 1977), 33–35 and 42.

11. Len Deighton, *Blood, Tears, and Folly* (London: Pimlico, 1995), 341–342.

12. Arnold, 242.

13. Unfortunately the military was not interested at that time, or for the duration of the war, in the work of Robert Goddard, the American rocket expert. In the second half of 1940, after a lengthy presentation by Goddard, General George Brett, chief of materiel in the air corps, wrote to Goddard: "The proposals as outlined in your letter . . . have been carefully reviewed. . . . While the Air Corps is deeply interested in the research work being carried out by your organization . . . it does not, at this time feel justified in obligating further funds for basic jet propulsion research and experimentation." The navy's reply was essentially similar. See Milton Lehman, *This High Man, The Life of Robert H. Goddard* (New York: Farrar, Straus, 1963), 310. When after the war German rocket scientists were interrogated by the Americans on technical points, one of them said, "Why don't you ask your own Dr. Goddard?" See Lehman, 389.

14. Toward the end of the war, this airplane was actually used as an interceptor against Allied bombers.

15. As will be shown in the case of the Soviet atomic bomb, this kind of sentiment can exist anywhere. It is another manifestation of the NIH syndrome but without even the "reasons" sometimes expressed by those afflicted with it.

16. Bekker, 486.

17. Buderi, 156–157.

18. This section is based to a large extent on David Holloway, "How

the Bomb Saved Soviet Physics," *The Bulletin of the Atomic Scientists* (November/December, 1994), 46–55.

19. This, incidentally, was the case also in the military in the twenties and thirties, when the "politruks"—in essence, political supervisors, most of whom had no military training—had the final say in the planning and execution of military operations, down to the company level. This aberration was abolished during the Second World War and only partially reinstituted afterward.

20. The Ukraine was considered Europe's grain storehouse and was an important part of the 1939 German-Soviet pact and later, part of the reason for the 1941 Nazi invasion.

21. Kurchatov was probably the first Russian who suspected that the United States was embarking on a nuclear program. In June of 1940 two of his colleagues published a scientific paper reporting on spontaneous fission in uranium. They were surprised at the lack of American reaction to what at the time was a rather important finding. Adding this to a sudden dearth of American publications in the field, the Soviets then concluded that the Americans were hiding a big secret about uranium. The physicists alerted their superiors to their suspicions and they in turn alerted Communist "sleepers" in the west.

22. Holloway, 54. Holloway also says that there was another version to the story: The physicist approached Beria and asked him to call the conference off, saying that it would harm Soviet physics and interfere with the atomic project. Beria replied that he could not make a decision on the matter and went to Stalin, who gave the reply quoted here but canceled the conference.

23. One cannot avoid wondering if on the whole the world would not have been a better place today if the "party line" had won that struggle too.

CHAPTER 8. TECHNOLOGICAL DEVELOPMENTS AND SCIENCE FICTION

1. In every development project, there are the "known unknowns" and the "unknown unknowns." The first are usually nonexisting items or technologies but that are considered to be possible to acquire or develop within reasonable time. The second are the nasty surprises, usually in the form of some unexpected physical phenomena, that affect otherwise per-

fectly good equipment or components. These may require total redesign of critical components or even developing new technologies, with all the attendant problems. Buderi (235–237) relates how some research (in the Rad Lab in Boston) on a centimeter wavelength radar was progressing nicely until the equipment started failing. It turned out that the initial work had been done in the winter, when the air was very dry. In the spring, air humidity increased and water vapor absorbed too much energy in that particular wavelength. This natural effect killed the project's value for the war effort.

2. See Guy Hartcup, *The Silent Revolution* (London: Brassey's, 1993), which deals with the period 1945–1985.

3. S. Glasstone and P. Dolan, "The Electromagnetic Pulse and Its Effects," in *The Effects of Nuclear Weapons* (Washington, DC: DOD & DOE, 1977), 516–525.

4. An excellent such book does exist, *War Day and the Journey Onward*, by Whitley Strieber and James W. Kunetka (New York: Holt, Rinehart and Winston, 1984). It describes the effects of EMP on people's everyday lives in the United States after it is generated during a short (and otherwise not very damaging) nuclear war between the Soviet Union and the U.S. In a similar vein, a short story that was published in the 1940s dealt with the accidental release of some gas that destroyed all paper on earth, and the difficulties of adjustment.

5. See, for example, chapter 5 about the difficulties German submarine crews experienced with the Naxos receivers, or to quote for a change "a happy ending," the introduction, in the nick of time, of British radar.

6. Norman R. Augustine, *Augustine's Laws* (New York: American Institute of Aeronautics and Astronautics, 1983), 123–124 and 175–176, respectively.

7. Price, *Aircraft versus Submarine*, 13.

8. Emphasizing the point of bringing the astronaut back is not an idle one. In December 1962, the prestigious American Institute of Aeronautics and Astronautics published in its monthly magazine a paper about a one-way mission to the Moon, in order to be the first to plant there the American flag. The technical justification for this rather unusual suggestion being that a one-way mission is much simpler to accomplish and thus can be carried out in a much shorter time. The paper, which on the technical side was detailed and accurate, envisioned the lone astronaut being resupplied by unmanned supply rockets until such time (presumably within a few years) that the United States could develop the capability to

bring the man back. See John M. Cord and Leonard M. Seale, "The One-Way Manned Space Mission," *Aerospace Engineering*, Vol. 21, No. 12 (December 1962): 60–61, 94–101. The paper was later turned into an excellent work of fiction. Hank Searle, *Project Pilgrim* (Greenwich, CT: Fawcett, 1964).

9. In a 1962 tape released by the late President John F. Kennedy's library in Boston in August 2001, the president tells James Webb, then head of NASA, that beating the Russians must be the goal, otherwise "we shouldn't be spending this kind of money, because I'm not that interested in space." Quoted by Robert F. Dorr in "Washington Watch," *Aerospace America*, Vol. 39, No. 10 (2001), 8.

10. Without going into technical explanations, it should be pointed out that there are many ways to produce laser beams, and they can be produced in a very wide range of wavelengths. There are even x-ray lasers.

CHAPTER 9. TECHNOLOGICAL SURPRISE AND TECHNOLOGICAL FAILURE

1. British intelligence took a dead man, dressed him as an officer, and put in his briefcase various letters and documents (actually written by high-ranking officers) implying that the landing in Sicily was a cover plan for landings in Sardinia and Greece. The bogus "Major Martin" was then eased from a submarine into the sea near Spain. His body washed ashore and copies of the papers were duly delivered to Berlin. The Germans swallowed the bait and moved troops and other resources to the "threatened" locations, thereby weakening defenses at the real landing points.

2. The British paratroopers ran into two unexpected German armored divisions. The planners knew about the arrival of these divisions but chose to disregard this information due to political considerations. See Cornelius Ryan, *A Bridge Too Far* (New York: Simon & Schuster, 1974), 157–160.

3. The Styx had a range of about forty kilometers compared with twenty kilometers for the Israeli Gabriel. The Israelis, who were aware of this discrepancy in range, could not do much about the range of the first-generation Gabriel that was developed from a ground-to-ground missile. They opted instead to use other means, with better all-around utility, to equal the odds. In the sea engagements in 1973, all the Styx missiles fired were made to miss until the Israeli missile boats closed the range and could fire their own missiles.

4. Samuel Eliot Morison, *History of United States Naval Operations in World War II, Vol. VI* (Boston: Little, Brown, 1949), 195–196.

5. Thomas G. Mahnken, "Gazing at the Sun: The Office of Naval Intelligence and Japanese Naval Innovation, 1918–1941," in *Intelligence and National Security*, Vol. 11, No. 3 (July 1996), Frank Cass, London, 432–433.

6. At some stage in the fighting in Europe, even the American resources were strained by the terrible losses in the tank forces. Tanks were taken from training units and shipped to Europe. See Ross, 279–280.

7. D'Olier, 7.

8. Emerson, 39.

9. Unless very well prepared or extremely lucky, solving such a problem and implementing it takes time, and in a fast-paced battle or with a resolute enemy, time will be short. See also chapter 4, about the fast reaction of the British to the initial gas attacks.

10. An interesting "might have" extension to the Hutier technique was an airborne assault and landing, or parachuting, of troops and machine guns behind enemy lines. Toward the end of the war it was technically possible to do this with reasonable numbers, and it was in fact suggested by General Billy Mitchell and endorsed by General Pershing. The war, however, ended before this could be tried, but a successful exercise was performed after the war at Kelly Field. See Roger Burlingame, *General Billy Mitchell* (New York: Signet, 1956), 81–82; and Arnold, 398.

11. F. W. Winterbotham, *The Ultra Secret* (New York: Harper & Row, 1974), 80.

12. It is actually the equivalent of those soldiers wearing a cloak conferring invisibility, which is even more "imaginary."

13. Azriel Lorber, "Active Defense—A New Concept for Submarine Defense," from the Proceedings of the 1992 Undersea Defense Technology Conference (London, 1992).

14. "Soft-kill" denotes the prevention of a hit by diverting the incoming weapon from its course.

BIBLIOGRAPHY

Air Ministry. *The Origins and Development of Operations Research in the Royal Air Force*. London: Air Ministry, Air Publication 3368, 1963.

Allison, David K. *New Eye for the Navy: The Origin of Radar at the Naval Research Laboratory*. Washington, DC: Naval Research Laboratory, NRL Report 8466, 1981.

Anderberg, Bengt, and Myron L. Wolbarsht. *Laser Weapons*. New York: Plenum, 1992.

Anon. "Los Buques Alemanes Internados en Nuestra Pais," *Revista de la Liga Maritima de Chile*, Ano 5, No. 31 (1919): 1–2.

Arnold, H. H. *Global Mission*. Blue Ridge Summit, PA: TAB Books, Military Classics Series, 1989.

Augustine, Norman R. *Augustine's Laws*. New York: American Institute of Aeronautics and Astronautics, 1983.

Barnes, G. M. *Weapons of World War II*. New York: D. Van Nostrand, 1947.

Baucom, Donald R. *The Origins of SDI, 1944–1983*. Lawrence: University Press of Kansas, 1992.

Baxter, James P. *Scientists against Time*. Boston: Little, Brown, 1946.

Bekker, Cajus. *The Luftwaffe War Diaries*. New York: Ballantine, 1971.

Bernstein, Jeremy, and David Cassidy. "Bomb Apologetics: Farm Hall, August 1945," *Physics Today*, Vol. 48, No. 8, Part 1 (1995) 32–36.

Bettenbender, John, and George Fleming, eds. *Famous Battles*. New York: Dell, 1970.

Blair, Clay Jr. *Silent Victory*. Philadelphia: J. B. Lippincott, 1975, Second Printing.

Borkin, Joseph. *The Crime and Punishment of I. G. Farben*. New York: The Free Press, 1978.

Bowen, E. G. *Radar Days*. Bristol, UK: Adam Hilger, 1987.

Boylan, Bernard L. "The Development of the American Long-Range Escort Fighter." Ph.D. diss., University of Missouri, 1955.

Bradley, Omar N. *A Soldier's Story*. New York: Henry Holt and Company, 1951.

Brickhill, Paul. *The Dam Busters*. New York: Ballantine, 1955.
Brodie, Bernard, and Fawn M. Brodie. *From Crossbow to H-Bomb*. Bloomington: Indiana University Press, 1973.
Brown, Michael E. *Flying Blind*. Ithaca, NY: Cornell University Press, 1992.
Buderi, Robert. *The Invention That Changed the World*. New York: Simon & Schuster, 1997.
Burlingame, Roger. *General Billy Mitchell*. New York: Signet, 1956.
Bush, Vannevar. *Modern Arms and Free Men*. New York: Simon & Schuster, 1949.
Caidin, Martin. *Flying Forts*. New York: Ballantine, 1968.
Carus, W. Seth. Testimony in *Report of the Commission to Assess the Ballistic Missile Threat to the United States*. Appendix III, Unclassified Working Papers. Pursuant to Public Law 201, 104th Congress, Washington, DC: July 15, 1998, 79–85.
Chant, Christopher, ed. *Warfare and the Third Reich*. New York: Smithmark, 1996.
Churchill, Winston S. *Closing the Ring*. Boston: Houghton Mifflin, 1949.
———. *The Hinge of Fate*. Boston: Houghton Mifflin, 1949.
———. *Their Finest Hour*. Boston: Houghton Mifflin, 1949.
———. *The World Crisis 1911–1914*. New York: Charles Scribner's Sons, 1924.
———. *The World Crisis 1915*. New York: Charles Scribner's Sons, 1923.
Clark, Ronald W. *War Winners*. London: Sidgwick & Jackson, 1979.
Clausewitz, Carl von. *On War*. New York: Barnes & Noble, 1966.
Coffey, Thomas M. *Iron Eagle, the Turbulent Life of General Curtis LeMay*. New York: Crown, 1986.
Cohen, Eliot A., and John Gooch. *Military Misfortunes*. New York: The Free Press, 1990.
Commission on the Organization of the Government for the Conduct of Foreign Policy. *Adequacy of Current Organization: Defense and Arms Control*. Volume 4, Appendix K, June 1975.
Cord, John M., and Leonard M. Seale. "The One-Way Manned Space Mission," *Aerospace Engineering*, Vol. 21, No. 12 (1962): 60–61 and 94–101.
David, Saul. *Military Blunders*. New York: Carroll & Graf, 1998.
Deighton, Len. *Blood, Tears, and Folly*. London: Pimlico, 1995.
———. *Fighter*. New York: Ballantine, 1977.
de Seversky, Alexander. *Victory through Air Power*. New York: Simon & Schuster, 1942.
Dixon, Norman. *On the Psychology of Military Incompetence*. New York: Basic Books, 1976.

D'Olier, Franklin. Chairman, *The United States Strategic Bombing Survey, Summary Report, (European War)*. September 1945.
Dornberger, Walter. *V-2, The Nazi Rocket Weapon*. New York: Ballantine, 1954.
Dorr, Robert F. "Washington Watch," *Aerospace America*, Vol. 39, No. 10 (2001): 8.
Dunnigan, James F. *Digital Soldiers*. New York: St. Martin's, 1996.
Dupuy, Trevor N. *The Evolution of Weapons and Warfare*. London: Jane's, 1980.
Ellis, John. *The Social History of the Machine Gun*. New York: Arno, 1981.
Emerson, William R. *Operation Pointblank (A Tale of Bombers and Fighters)*. U.S. Air Force Academy: The Harmon Memorial Lectures in Military History, 1962.
Emme, Eugene M. *The Impact of Air Power*. New York: D. Van Nostrand, 1959.
Everitt, Don. *K-Boats, Steam-Powered Submarines in World War I*. Annapolis, MD: Naval Institute Press, 1999.
Fisher, David E. *A Race on the Edge of Time*. New York: McGraw-Hill, 1988.
Ford, Brian J. *Allied Secret Weapons*. New York: Ballantine, 1971.
Freeman, Dyson. *Disturbing the Universe*. New York: Harper & Row, 1979.
Friedrich Der Grosse. *Die General-Principia Vom Kriege 1753*. (Hebrew Translation), Tel Aviv, Israel: Ma'aracoth, IMOD, 1979.
Fuller, J. F. C. *Armament and History*. New York: Charles Scribner's Sons, 1945.
Galland, Adolf. *The First and the Last*. New York: Ballantine, 1954.
Gannon, Robert. *Hellions of the Deep*. College Park: Pennsylvania State University Press, 1996.
Garden, Timothy. *The Technology Trap*. London: Brassey's, 1989.
Glasstone, S., and P. Dolan. "The Electromagnetic Pulse and Its Effects," in *The Effects of Nuclear Weapons*. Washington, DC: DOD & DOE, 1977.
Gray, Edwyn. *The Devil's Device*. Annapolis, MD: Naval Institute Press, 1991.
Grey, C. G. *The Luftwaffe*. London: Faber and Faber, 1944.
Grierson, John. *Jet Flight*. London: Sampson Low, Marston, 1945.
Groves, Leslie. *Now It Can Be Told*. New York: Harper & Row, 1962.
Guderian, Heinz. *Panzer Leader*. New York: Ballantine, 1961.
Gunston, Bill. *Submarines in Color*. New York: Arco, 1977.
Handel, Michael I. "Clausewitz in the Age of Technology," in *Clausewitz and Modern Strategy*. London: Frank Cass, 1986.

Hansen, Chuck. *U.S. Nuclear Weapons*. Arlington, TX: Aerofax, 1988.

Harford, James. *Korolev*. New York: John Wiley & Sons, 1997.

Hartcup, Guy. *The Challenge of War*. Newton Abbot, UK: David & Charles, 1970.

———. *The Silent Revolution*. London: Brassey's, 1993.

———. *The War of Invention*. London: Brassey's, 1988.

Hogg, Ian V. *The Complete Machine Gun, 1885 to the Present*. London: Phoebus, 1979.

Holloway, David. "How the Bomb Saved Soviet Physics," *The Bulletin of the Atomic Scientists* (November/December 1994): 46–55.

Hutcheon, Wallace S. *Robert Fulton*. Annapolis, MD: Naval Institute Press, 1981.

Irving, David. *The German Atomic Bomb*. New York: Simon & Schuster, 1967.

Jablonski, Edward. *Flying Fortress*. Garden City, NY: Doubleday, 1965.

Johnson, Brian. *The Secret War*. New York: Methuen, 1978.

Johnson, Clarence "Kelly" L., with Maggie Smith. *Kelly*. Washington, DC: Smithsonian Institution Press, 1985.

Johnson, Clint. *Civil War Blunders*. Winston-Salem, NC: John F. Blair, 1997.

Johnson, David E. *Fast Tanks and Heavy Bombers, Innovation in the U.S. Army, 1917–1945*. Ithaca, NY: Cornell University Press, 1998.

Johnson, Melvin M., and Charles T. Haven. *Automatic Weapons of the World*. New York: William Morrow, 1945.

Jones, Archer. *The Art of War in the Western World*. Urbana, IL: University of Illinois Press, 1987.

Jones, R. V. *Reflections on Intelligence*. London: Heinemann, 1989.

———. *The Wizard War*. New York: Coward, McCann & Geoghegan, 1978.

Kahn, David. *The Code Breakers*. New York: Scribner, 1996.

Keaney, Thomas A., and Eliot A. Cohen. *Revolution in Warfare? Air Power in the Persian Gulf*. Annapolis, MD: Naval Institute Press, 1995.

Keegan, John. *The Face of Battle*. New York: Penguin, 1978.

———. *The Second World War*. London: Viking, 1990.

Keegan, John, and Richard Holmes. *Soldiers, A History of Men in Battle*. London: Hamish Hamilton Ltd., 1985.

Kelsey, Benjamin S. *The Dragon's Teeth?* Washington, DC: Smithsonian Institution Press, 1982.

Kennedy, Paul M. *The Rise and Fall of British Naval Mastery*. London: Ashfield Press, 1983.

Ketchum, Richard M. *The Borrowed Years 1938–1941*. New York: Random House, 1989.

Kirkpatrick, D. L. I. "Do Lanchester's Equations Adequately Model Real Battles?" *RUSI Journal* (June 1985): 25–27.

Laffin, John. *Brassey's Battles*. London: Brassey's, 1995.

Lehman, Milton. *This High Man, The Life of Robert H. Goddard*. New York: Farrar, Straus, 1963.

Liddel Hart, Basil H. *Foch, the Man of Orleans*. Boston: Little, Brown, 1932.

Lorber, Azriel. "Active Defense—A New Concept for Submarine Defense," Proceedings of the 1992 Undersea Defense Technology Conference. London (June 30–July 2, 1992): 261–266.

MacIntyre, Donald, and Basil W. Bathe. *Man-of-War*. New York: Castle, 1974.

Macksey, Kenneth. *Military Errors of World War Two*. London: Cassell, 2000.

Mahnken, Thomas G. "Gazing at the Sun: The Office of Naval Intelligence and Japanese Naval Innovation, 1918–1941," *Intelligence and National Security*, Vol. 11, No. 3 (1996): 424–441.

Matson, Wayne R. *Cosmonautics*. Washington, DC: Cosmos, 1994.

Mayo, Lida. "The Ordnance Department: On Beachhead and Battlefront," in *United States Army in World War II—The Technical Services*. Washington, DC: Office of the Chief of Military History, United States Army, 1968.

McCue, Brian. *U-Boats in the Bay of Biscay*. Washington, DC: National Defense University Press, 1990.

Miller, Russell. *The Soviet Air Force at War*. Alexandria, VA: Time-Life, 1985.

Millett, Allan R., and Williamson Murray. *Military Effectiveness (Volume 1, The First World War; Volume 2, The Interwar Period; Volume 3, The Second World War)*. Boston: Unwin Hyman, 1988.

Mordal, Jacques. *25 Centuries of Sea Power*. New York: Clarkson N. Potter, 1965.

Morison, Elting E. "Gunfire at Sea: A Case Study of Innovation," in *Men, Machines, and Modern Times*. Cambridge, MA: MIT Press (1966), 17–44.

Morison, Samuel E. *History of United States Naval Operations in World War II, Vol. VI*. Boston: Little, Brown, 1949.

Morris, Donald R. *The Washing of the Spears*. New York: Simon & Schuster, 1965, fourth printing.

Morse, Philip M., and George E. Kimball. *Methods of Operations Research*. OEG Report No. 54, Operations Evaluation Group, Office of the CNO. Washington, DC: Navy Department, 1946.

Murray, Williamson. *The Luftwaffe, 1933–1945: Strategy for Defeat*. Washington, DC: Brassey's, 1996.

Murray, Williamson, and Allan R. Millett, eds. *Military Innovation in the Interwar Period*. Cambridge, UK: Cambridge University Press, 1996.

Na'aman, Shlomo, and Roni Cohen. "Tank Armament, Past, Present and Future," (in Hebrew), *Ma'arachot* No. 247–248 (1975): 22–54.

Norris, Robert S. "Estimated U.S. and Soviet/Russian Nuclear Stockpiles, 1945–1994," *The Bulletin of the Atomic Scientists* (November/December 1994).

O'Donnell, James P. "The Secret Fight That Doomed the Luftwaffe," *Saturday Evening Post* (April 8, 1950): 22–23, 100, 102–105.

Overy, Richard. *Why the Allies Won*. New York: W. W. Norton, 1995.

Pacific War Research Society, ed. *Japan's Longest Day*. New York: Ballantine, 1983.

Pawle, Gerald. *Secret Weapons of World War II*. New York: Ballantine, 1968.

Peebles, Curtis. *Dark Eagles*. Novato, CA: Presidio, 1995.

Postan, M. M., D. Hay, and J. D. Scott. *Design and Development of Weapons—Studies in Government and Industrial Organisation*. London: Her Majesty's Stationery Office and Longmans, Green and Co., 1964.

Powers, Thomas. *Heisenberg's War*. Boston: Little, Brown, 1993.

Price, Alfred. *Aircraft versus Submarine*. London: Jane's, 1980.

———. *Instruments of Darkness*. London: Macdonald and Jane's, 1977.

Pritchard, David. *The Radar War*. Wellingborough, UK: Patrick Stephens Ltd., 1989.

Regan, Geoffrey. *The Guinness Book of Military Blunders*. Enfield, UK: Guinness Publishing, 1995.

Rhodes, Richard. *The Making of the Atomic Bomb*. New York: Penguin, 1988.

Rich, Ben R., and Leo Janos. *Skunk Works*. Boston: Little, Brown, 1994.

Ropp, Theodore. *War in the Modern World*. New York: Collier, 1974.

Rosen, Peter S. *Winning the Next War*. Ithaca, NY: Cornell University Press, 1991.

Rosenberg, Bruce A. *Custer and the Epic of Defeat*. University Park: Pennsylvania State University Press, 1974.

Ross, G. Macleod (in collaboration with Sir Campbell Clark). *The Business of Tanks, 1933 to 1945*. Ilfracombe, UK: Arthur H. Stockwell Ltd., 1976.

Ryan, Cornelius. *A Bridge Too Far*. New York: Simon & Schuster, 1974.

Sapolsky, Harvey M. *The Polaris System Development*. Cambridge, MA: Harvard University Press, 1972.

Seagrave, Sterling. *Yellow Rain*. New York: M. Evans, 1981.

Searle, Hank. *Project Pilgrim*. Greenwich, CT: Fawcett, 1964.

Shafritz, Jay M. *Words on War*. New York: Prentice Hall, 1990.

Speer, Albert. *Inside the Third Reich*. New York: Macmillan, 1970.

Steinhoff, Johannes. *The Final Hours*. Baltimore, MD: Nautical & Aviation Publishing Company of America, 1977.

Sternhell, Charles M., and Alan M. Thorndike. *Antisubmarine Warfare in World War Two*. OEG Report No. 51, Operations Evaluation Group, Office of the CNO. Washington, DC: Navy Department, 1946.

Strieber, Whitley, and James W. Kunetka. *War Day and the Journey Onward*. New York: Holt, Rinehart and Winston, 1984.

Swords, S. S. *Technical History of the Beginnings of Radar*. London: Peter Peregrinus Ltd., 1986.

Tompkins, John S. *The Weapons of World War III*. London: Robert Hale Ltd., 1966.

Townsend, Peter. *Duel of Eagles*. New York: Simon & Schuster, 1970.

Tuchman, Barbara W. *A Distant Mirror, The Calamitous 14th Century*. London: Macmillan, 1979.

Van Crefeld, Martin. *Technology and War*. New York: The Free Press, 1989.

Watson-Watt, Robert. *Three Steps to Victory*. London: Odhams Press Limited, 1957.

Watts, Anthony J. *The Royal Navy, An Illustrated History*. Annapolis, MD: Naval Institute Press, 1994.

Welchman, Gordon. *The Hut Six Story*. New York: McGraw-Hill, 1982.

Whiting, Kenneth R. *Soviet Air Power*. Boulder, CO: Westview, 1986.

Whittle, Frank. *Jet*. London: Pan Books, 1957.

Winterbotham, F. W. *The Ultra Secret*. New York: Harper & Row, 1974.

Yergin, Daniel. *The Prize*. New York: Simon & Schuster, 1992.

Zaloga, Steven J. "Polish Cavalry against the Panzers," *Armor Magazine* (January–February 1984): 26–31.

INDEX

ABOUT THE AUTHOR

Azriel Lorber is an aerospace engineer and a retired Israeli Defense Force officer. Born in Poland in 1935, he and his family emigrated shortly before the German invasion, arriving in British-controlled Palestine in July 1939. He studied engineering in the United States, receiving his Ph.D. from Virginia Tech in 1974. After working most of his career in research and development in the Israeli defense industry, he participated as a visiting scholar in the Security Studies Program at the Massachusetts Institute of Technology in 1998–1999. Lorber is currently an independent consultant and lecturer. He has written widely in numerous scientific, technical, and military journals and is the author or editor of five books, including three published in Hebrew, as well as *Theater Ballistic Missile Defense* (2001). *Misguided Weapons* is his latest book.